So, You Bought A Baofeng Radio...

Now what?

Rodney Biddle
Ham Radio Operator KX4H, AE, VE
Ham Radio Instructor, Testing Coordinator

http://www.SoYouBoughtABaofeng.com/

So, You Bought a Baofeng Radio – Now what?

Table of Contents

Forward 4
Introduction 6
Chapter 1: Legalities First 8
Chapter 2: Getting to know your Radio 16
Chapter 3: Navigating the Radio MENU system 30
Chapter 4: REAL Radio Capabilities 48
Chapter 5: Understanding VHF & UHF 56
Chapter 6: Power and your Radio 60
Chapter 7: Antennas 66
Chapter 8: What Frequency do I use? 76
Chapter 9: Talk About Repeaters 82
Chapter 10: Radio Nets and Finding them 100
Chapter 11: Programming with CHIRP 106
Chapter 12: Hand programming the Baofeng 119
Chapter 13: Lingo & Talking on the Radio 129
Chapter 14: Finding a Club 133
Chapter 15: Basics About Coax & Your Baofeng 135
Chapter 16: Are you going to Use that Radio? 139
Chapter 17: Advanced Topics 141
Chapter 18: Non-Ham Communications 147
Chapter 19: Getting Licensed & Legal 151
Chapter 20: Some recommended Accessories 157
Appendix A: Calling Frequencies 165
Appendix B: Bands for LEGAL Use 165
Appendix C: GMRS/FRS Frequencies & Data 166
Appendix D: MURS Frequencies 167

Appendix E: VHF Marine Frequencies 169
Appendix F: Frequency Coordinators 171
Appendix G: Website References 176
Appendix H: SWR Meters 177
Appendix I: Parts Lists, Links to parts, Sources 180
Appendix J: CTCSS Squelch Tones (Hz) 182
Appendix K: DCS Codes 182
Appendix L: Phonetic Alphabet 183
Appendix M: Radio Horizon Antenna Heights 184
Appendix N: Weather Radio Frequencies 185
Appendix N: Baofeng Menu Settings/Values 186
Appendix O: Chirp Radio Settings Menus (Baofeng Radios) 193

So, You Bought a Baofeng Radio – Now what?

So, You Bought a Baofeng Radio – Now what?

So, You Bought a Baofeng Radio – Now what?

Forward

My first actual Ham radio was a Baofeng UV-5r. I had dabbled with amateur radio years ago in college and am actually a holder of a radio license from college still. But I had not jumped into Ham Radio until after I started dabbling in the Prepper life.

In considering what life would be like if there were a disaster of epic proportions, I started thinking about technology that I would like to know. I am an engineer by trade and have been working in the Computer industry for over twenty years. I have worked and supported large world-wide data centers with so many dependencies and critical points of failures, the likelihood of those systems remaining up and fully intact after an EMP or CME disaster would be extremely unlikely – Not for a while at least.

Ham radio is a science and an art. One that has been around for over a hundred years, and in some ways has changed little. The radios have become smaller and portable, but the signaling and analog systems have stayed very similar for decades. While the computer and data center industries seem to revolutionize within just years, so much about Ham Radio stays similar over the years. I guess this also may have been a big draw for me to the art.

I am an educator. Part of my learning process is to study and learn, to write, document and to pass on what I have learned. When I got my first Baofeng, the manual was so awful it didn't help much. I struggled with the radio, studied and got my Tech license. But even with the tech license there was a lot to learn. I watched YouTube videos, bought toys to play with for the radio such as antennas, and I even broke down and bought a mini simplex repeater which allowed me to do some solo range testing with the radios. Hooking up one radio to the repeater and hiking

So, You Bought a Baofeng Radio – Now what?

out testing the range based on different settings and antennas I was using.

There was so much I was trying to learn which is part of what inspired me to write this book. I wanted to put a resource together that I could say should be a part of every prepper's kit. A resource for those just getting into the art and to help side-step some of the confusion with using of your first radio.

So, You Bought a Baofeng Radio – Now what?

Introduction

You just bought your first Baofeng Radio!

If you are purchasing this book as well, you probably fall into one of these categories:

- You are Brand New to Ham Radio and you are using this as your First starter Radio. If so – Wonderful. The Baofeng is a great first radio to start off with.

- You are a Prepper or participate in some type of Prepping community and the Baofeng Radio is the preppers Radio of Choice. But – Before you toss the radio in your Faraday box for someday when you might need it – You need to learn some things first about the radio. Otherwise you won't be able to use is when you need to.

- You have purchased the radio for a spare or backup radio (If you're an experienced ham you probably don't need this book unless your looking for a reference to hand programming the radio).

- Or you have purchased the radio for use as a GMRS/FRS/MURS radio for hunting, camping, 4-wheeling or other activity and you want an inexpensive radio (Make sure to closely read Chapter 1 in this case).

The Baofeng is a Chinese manufactured radio that works on a wide range of frequencies in the VHF and UHF radio bands. In fact, the range of frequencies goes beyond the frequencies that the radio is legal for use with here in the United States. Yes – The radio is legal to use for some operations but be aware that improper use of the radio can put you into hot water so please read chapter 1 carefully where we can review these legalities with you.

So, You Bought a Baofeng Radio – Now what?

These radios are generally based on the older UV-5r 5-watt handheld radio which came out several years ago. Since then there have been several models which we will review in this book. Hobbyists, recreational enthusiasts, Ham operators, Hunters and Preppers all use this radio as a starter radio, a backup radio, or for some even an emergency radio which many will throw into a box for a SHTF scenario.

While the radio is a great little starter to use and learn with – it is not what many think it is. After all – you can pick these up for under $30 and to many they are just a cheap toss-away radio or when the radio does break.

For whatever reason you have purchased the radio – this book is intended to be a supplemental to help clarify many of the mis-notions about the radio and to help you get optimal use out of it. I myself own several, though I have also graduated to better and more powerful radios that I use on a frequent basis. But I can still hit the repeater network by me here in Florida, and through that repeater network and from my handheld radio I can communicate with other hams hundreds of miles away throughout Florida. As an avid hiker I always carry a radio with me on remote hikes – I have been in numerous areas with no cell coverage, but I am able to talk and listen to others on my little handheld.

So, let's get started – Welcome to the Baofeng world and congratulations on your new radio. Let's get you familiar with the radio and how to use is, and in particular let's get you knowing the pitfalls to avoid and understanding the proper legal use of the radio. And just as importantly if you decide to break the rules – let's make sure you know what those rules are.

So, You Bought a Baofeng Radio – Now what?

Chapter 1: Legalities First

This is the first chapter because this is the one area that most people don't have a tight grasp on when starting to work with the Baofeng radios. These radios work in such a wide range of frequencies and can communicate with so many radio systems that many assume that use with these systems is legal.

Well let's start by clarifying exactly what these radios are legal for. First – We are talking broadly about most of these radios. There are some very specific models imported, modified and sold by Baofeng Tech out of South Dakota that are legal for use beyond what we will talk about at first, but we will discuss that later in this chapter.

These radios can work within the following frequency ranges:

Radio Band	Radio Range	Legal HAM Range
VHF Band (2-Meter)	136 – 174 MHZ	144 – 148 MHZ
UHF Band (70-CM)	400 – 520 MHZ	420 – 450 MHZ

What this table shows us is that the radio itself is capable to transmit in frequencies that are outside of the frequency range for HAM radio. At first this sounds great – But here are the points you MUST understand:

1. For use with HAM Radio frequencies, the Radio Operator is considered the licensed component and NOT the radio. This allows legal use of these inexpensive radios by HAM Operators.

2. For using with all other radio systems, the radio MUST carry an FCC Certification. This radio does not carry a certification for other uses.

So, You Bought a Baofeng Radio – Now what?

3. The FCC Part 97 certification that the OPERATOR is certified under allows the use of these radios. But to use the radio on GMRS, MURS, VHF Marine or Commercial you will need a proper FCC Certification – or you are illegally communicating.

4. As of September 30, 2019, it is illegal for any radio to be sold (Including the Baofeng Radios) which can operate in the FRS band or other bands that they are not specifically licensed for.

5. To use the radios for GMRS, MURS or FRS the radio MUST carry an FCC Part 95 certification. Baofeng Tech has specialty radios that do carry this certification but are limited to working on specific frequency ranges for two-way communication. (More later)

So, what happens if you are using the radio in the wrong way? Well these radios are generally low power, so if you are using them and you are not around anyone or anywhere that you can cause interference probably nothing. Though if caught you may face fines of several thousands of dollars.

Where you will have problems is where you are creating "Harmful Interference" where your use of the radio is creating a problem and you are identified and reported. For instance – If you are using your radio and you interfere with the legal use of radios within range of your transmission you may have a problem. Many other legal uses of the frequencies that the Baofengs are capable of transmitting include private businesses (McDonald's, Burger King headsets, Schools, Business Radio, Emergency use, etc). IF YOU INTERFERE WITH THESE SYSTEMS YOU ARE BREAKING THE LAW.

Again though – Out in the woods hunting or camping, you are not likely to create interference through you are still breaking the law.

So, You Bought a Baofeng Radio – Now what?

CLARIFYING LEGAL USE
To clarify again, legal use of these radios includes use by a licensed ham radio operator within the frequencies as listed in the table on the prior page that have been set aside for ham radio operators. Any use outside of these frequencies or by a non-licensed ham operator is prohibited by the FCC.

LEGAL GMRS & MURS RADIO SOLUTIONS
What radios are available for legal GMRS & MURS Operation from Baofeng? Well, from Baofeng China none. But – From Baofeng Tech, who is the US Distributor for Baofeng Radios in the United States, they sell a GMRS-V1 and a MURS-V1 radio which is a Baofeng radio in different clothing. Well – same outside, mostly the same radio, but modified firmware. These radios can still be programmed for any frequency within the range of the Baofeng radio for receiving, but for transmitting they are locked to only work within the frequencies for each radio service.

The GMRS-V1 is a 5-watt radio with a replaceable antenna while the MURS-V1 is limited to 2-watts (Per MURS Rules) but also with a replicable antenna. Both radios also carry the appropriate FCC Part 95 certification for use on their respective radio services.

LEGAL COMMERCIAL RADIO
Just like having a solution for GMRS & V1, Baofeng Tech has also developed a UV-82C radio which comes with the appropriate Part 90 FCC approval required for commercial radio use. The requirement for this approval is that the radio has VFO locked out and has modifications so that the radio must be programmed via software only. Once programmed however, some of the limitations imposed on the radio out of the box can be turned off. If you are looking to use the radio for commercial use (And you have a commercial radio use license), this will

So, You Bought a Baofeng Radio – Now what?

be the radio to purchase. At a price of less than $60 these radios are a bargain.

WHAT ABOUT JUST LISTENING AND MONITORING?

We have provided a description of legal use at this point. But what if you just want to monitor a channel? These radios are COMPLETELY legal for listening on and monitoring. And – If you are in a life-threatening emergency, then the FCC does allow for use of these radios without a license.

And what about if a SHTF scenario occurs? This is the scenario that many preppers call the worst case. Some type of breakdown – possibly power loss, could be civil unrest, or the collapse of society in general. Well, of course if there is this level of collapse there likely won't be anyone going around to fine folks for illegal transmission of a radio. In these cases, we will all have plenty of other worries to be concerned about.

Baofeng Radios and Spurious Emission problems

Here is an area that the Baofeng Radios are notorious for – Spurious Emissions. So much so that they can produce illegal transmissions in frequencies that are outside of the frequencies for Ham Operators to use.

So – What are Spurious Emissions?

This is when the radio is actually broadcasting not only on the frequency that you intend to transmit on, but also on additional frequencies. The joke about Baofeng radios (Or other cheap Chinese radios) is "Why buy a radio that can only transmit on one frequency when you can transmit on several.".

So, You Bought a Baofeng Radio – Now what?

What happens is as the radio is transmitting out at a high power on a primary frequency, the filtering in the radio does not do an adequate job of filtering out harmonic frequencies and the radio actually transmits on a second, third or even fourth harmonic frequency other than your intended frequency and at a power above what they should be transmitting at. In the image below you can see the high peak about the green line which indicates the transmit power. But here you will also see additional peaks above the green line power-level which should either exist, or at least not exist above the green power level.

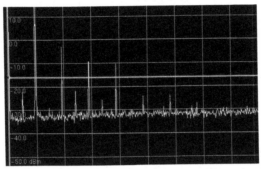

A Frequency Sampling from a poorly behaving Baofeng shows harmonic transmissions at power levels above appropriate levels indicating poor filtering.

This indicates that this particular radio is out-of-spec per the FCC rules and should not be legal for use. This is probably the number one argument against these radios – Improper filtering on most will create these additional transmissions that could cause noise or interference to legal use of other radio-type of equipment – Harmful emissions. The failure rate among Baofeng Radios does tend to be high also.

This still doesn't stop the popularity of this radio and use of the radio by Hams and Non-Hams alike. I recently participated in a small gathering of Ham Radio students going over setting up and programming of radios using Chirp. The only restrictions we placed was that only licensed Ham

So, You Bought a Baofeng Radio – Now what?

Operators could have frequencies programmed for two-way communication where all non-licensed Hams would have their radios programmed for listening only – No transmit. In the class, over 90% of the radios were Baofeng models, and most of those were BF-F8HP model radio which are popular in the Prepper/Readiness communities.

- https://hamgear.wordpress.com/2013/01/30/baofeng-uv-5r-spectrum-analysis-revisited/

- https://hackaday.com/2016/12/14/measuring-spurious-emissions-of-cheap-handheld-transceivers/

- https://hamradiohawaii.wordpress.com/2018/03/24/most-baofeng-and-a-few-wauxon-do-not-comply-with-part-97-standards/

Following is the statement in the Baofeng Owner's Manual regarding licensed legal use:

FRS, GMRS, MURS, PMR446
You may be tempted to use FRS, GMRS, MURS (in the USA) or PMR446 (in Europe) frequencies. Do note however that there are restrictions on these bands that make this transceiver illegal for use.

So, You Bought a Baofeng Radio – Now what?

Following is the FCC Notice that is found in the Baofeng Manuals:

This equipment has been tested and found to comply with the limits for a Class B digital device, pursuant to part 15 of the FCC Rules. These limits are designed to provide reasonable protection against harmful interference in a residential installation. This equipment generates uses and can radiate radio frequency energy and, if not installed and used in accordance with the instructions, may cause harmful interference to radio communications. However, there is no guarantee that interference will not occur in a particular installation. If this equipment does cause harmful interference to radio or television reception, which can be determined by turning the equipment off and on, the user is encouraged to try to correct the interference by one or more of the following measures:

- *Reorient or relocate the receiving antenna.*
- *Increase the separation between the equipment and receiver.*
- *Connect the equipment into an outlet on a circuit different from that to which the receiver is connected.*
- *Consult the dealer or an experienced radio/TV technician for help*

Changes or modifications not expressly approved by the party responsible for compliance could void the user's authority to operate the equipment. This device complies with Part 15 of the FCC Rules. Operation is subject to the following two conditions:

    (1) This device may not cause harmful interference, and
    (2) This device must accept any interference received, including interference that may cause undesired operation.

WARNING: MODIFICATION OF THIS DEVICE TO RECEIVE CELLULAR RADIOTELEPHONE SERVICE SIGNALS IS PROHIBIITED UNDER FCC RULES AND FEDERRAL LAW.

So, You Bought a Baofeng Radio – Now what?

Chapter 2: Getting to know your Radio

Now it is time to start getting familiar with your radio. For the most part, and for the purposes of this book, there are two styles of the radios that we will be looking at. Both of these models can be considered "Families" of radios in which the controls are nearly identical as well as the batteries.

The classic Baofeng is based on the older UV-5R style radio. Several updated models use the same controls and same overall design with few variances in the firmware. What we are going to focus on for this family stye of radio is the newer 7/8-watt BF-F8HP radio which is a 7/8-watt handheld radio that uses the same cradle, batteries, and many accessories.

In addition, we are also going to be looking at the UV-82 style of radios. These are newer radios with updated firmware, an updated battery style, slightly different keypad, and a bit more rugged of a feel as shown in the photo below.

The Baofeng BH-F8HP on the left, the UV-82HP on the right. Note the radio on the right has an optional SMA-to-BNC Antenna adapter on it to simplify the switching between antennas. Both come with SMA-Antenna bases.

So, You Bought a Baofeng Radio – Now what?

Functionally both families of radios are equal in capabilities. Here are the general specifications for both specific models:

- Both radios are 7/8 watts of power (7-Watts UHF, 8-Watts VHF)
- Both have the same frequency range.
- Both support 128 channels for programming
- Both radios include LED Flashlights built in
- Both radios include the ability to receive FM Radio
- Both radios come fixed firmware – They cannot be updated.
- Both radios are available in multiple colors – The RED radio in this photo is my radio pre-programmed for ARES Emergency use.
- Both radios have Dual-Monitor capabilities (Channel-A & Channel-B)
- Both radios have a SMA replaceable antenna
- Both radios support the same dual-pin Kenwood-type connector for programming cable and microphones.

And here are the major differences between the radios:

- UV-82HP Radio comes with Dual PTT Buttons for Channel A & B
- UV=82 has a Slightly better Antenna
- UV-82 has 4-Tone burst setting
- UV-82 has a Larger speaker at 1000mW, the BF-F8HP speaker is a smaller 700mW speaker.
- UV-82 has a Smaller battery @ 1800mAh (Though larger are available)
- UV-82 is a Larger radio – Many like the feel of this radio better as I do.
- BF-F8HP Radio gas an VFO/MR Button for switching between Frequency and Channel modes.

So, You Bought a Baofeng Radio – Now what?

- The UV-82HP Radios require you to power off the radio, then while powering it back on hold down the [MENU] button while turning the radio on (Only a very minor inconvenience).
- Keypad layouts are different – Again only a very minor issue.
- The UV-82HP supports an "R-Tone" option for repeaters – Used to activate some repeaters.
- Each radio represents a different radio family which utilizes different batteries and cradles.
- The UV-82 series support the disabling of the VFO for commercial radio capabilities of the UV-82C Radio.

The Biggest Differences between the radios are...

For preppers, the biggest difference that preppers look at with the radios are the batteries. Being able to have multiple radios that use the same batteries and then radio users being able to share batteries is considered a strong convenience. Both radios have high-capacity batteries available to them

The radios also utilize different cradles for charging. I own a BF-F8HP Radio and a UV-5R radio, and both radios can use the same batteries, the same cradle, and even the same expansion batteries and battery eliminators.

Likewise, with the UV-82HP Radio I have, I can interchange batteries with my GMRS-V1 radio and use the same cradles. This goes for the other model radios which use the same form format including the UV-82C (Commercial), and the MURS-V1 radio. The BF-F8HP Radio uses the same battery and cradle as the older UV-5R Radios and the newer UV-5X3 Triband radio.

So, You Bought a Baofeng Radio – Now what?

What about Other Models?

Yes – Baofeng does have other model radios. But for emergency preparedness, these are the two I would recommend. First of all – you need to make sure you get a radio that may be able to share accessories with other radios. Baofeng produces some models where this is simply not the case.

I am an avid hiker – and when I found out Baofeng had a waterproof model a few years ago I snapped one up. I purchased the UV-82WP model. But – I did not get the right battery – so I purchased a spare battery later for it based on the battery part number. The right part number came in, wrong battery. I purchased a second after talking to the vendor, and again the wrong battery, right battery part number. Then I found out that there was ANOTHER waterproof model, so I ordered it. The GT-3WP is another model radio that uses the same part # for the battery, but a slightly different molded battery.

I also own a GT-3 Radio (Not waterproof) that uses *Another* battery. So, among the seven Baofeng radios I own, I have four different types of batteries – and some are near impossible to find. But – The BF-F8HP and the UV-82HP batteries are readily available and in multiple capacity sizes.

More recently Baofeng has released a newer 16/20-watt handheld radios. What I have heard is that these require ANOTHER type of batteries which are difficult to find. Another example of why I recommend sticking with specific models.

ABOUT THESE RADIOS & GETTING TO KNOW THEM

Both radios are very similar in operation, and across models in the same families they are almost identical. The notes on the following

So, You Bought a Baofeng Radio – Now what?

pages about each style radio will mostly carry across between the radios with slight differences. We will cover the controls of each radio and then jump into some of the most useful information for general operation of the radios.

Getting to know your UV-82HP Radio (And similar models)
Here is a quick run-down of the buttons and controls on the UV-82HP Radio. This also applies to the UV-82C, GMRS-V1, and MURS-V1 Radios.

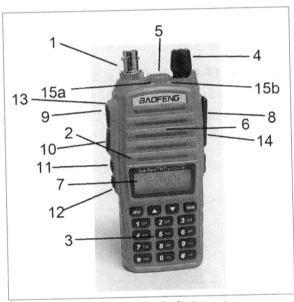

Baofeng UV-82 Radio Layout

(1) Antenna Connector (This radio has a BNC-Adapter I added for convenience when switching between antennas)
(2) LCD Radio Display
(3) Radio Keypad – Notice the layout is different from the BF-F8HP
(4) Power/Volume control knob
(5) LED Flashlight (Small powered light, but yes, convenient)
(6) 1000mW Speaker
(7) Microphone Location

So, You Bought a Baofeng Radio – Now what?

(8) Battery Release latch (On the back of the radio)'
(9) PTT-A Key – Push-To-Talk for Channel A Only
(10) PTT-B Key – Push-To-Talk for Channel B Only
(11) Side Key 1/[F] – Short-press Activates/Deactivates FM Radio mode. Long press to trigger the alarm functions.
(12) Side key 2/[M] – Press quickly to activate the flashlight in steady-on mode. Press quickly again to activate flashing flashlight. Press a third time to turn the flashlight off. Press and hold to turn squelch on the radio to OFF while the button is held.
(13) Strap buckle (Not shown)
(14) Accessory Jack for headset, microphone, or programming cable. This uses a 2-pin KENWOOD compatible connector. *Note – Some accessories such as Microphones may have difficulty fitting.*
(15) Statius LEDs – Green when receiving signal, red when transmitting.

Getting to know your BF-F8HP Radio (And similar models)

Here is a quick run-down of the buttons and controls on the BF-F8HP Radio. This also applies to the older UV-5R and newer UV-5X3 radios.

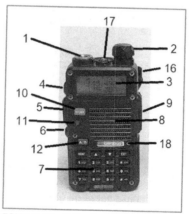

Baofeng BF-F8HP 8-Watt Radio based on the UV-5R Design

(1) Antenna Connector with the SMA Antenna Port
(2) Power/Volume knob
(3) Two-Line Display LCD

So, You Bought a Baofeng Radio – Now what?

(4) Call Key (Transmits an alarm code over the radio – Can be annoying)
(5) PTT Key (Single PTT Key on this radio)
(6) Monitor Key (Disables Squelch so you can hear weak signals)
(7) Keypad (Notice the different layout from the UV-82 series)
(8) Speaker & Microphone
(9) Accessory Jack (Programming cable, Microphone, Headset)
(10) VFO/MR Mode Key *(Not present on the UV-82 Series. On those radios you change mode by holding the [MENU] button while powering on the radio.)*
(11) Status LED
(12) A/B Select – Selects channel A or channel B
(13) Battery pack – On the back of the radio and not shown
(14) Battery contacts – Back-Bottom of the radio and not shown
(15) Battery release latch – At the TOP of the battery and not shown
(16) Lanyard loop
(17) LED Flashlight – Surprisingly useful, though not bright.
(18) BAND Button (Only on some UV-5R type models)

BF-F8HP Status LED (11)

The status LED on the front of the radio lights up GREEN when receiving a signal, RED when transmitting.

BF-F8HP Side Key 1 for CALL and Broadcast FM (4)

Your little Baofeng has an FM Radio in it. Not the best quality radio, but handy and convenient. The radio will play the FM station of your choosing until a radio caller comes onto the frequency you are monitoring, then when the frequency clears will start playing again. Pressing the [CALL] button briefly and releasing it activates the radio.

This button also has a second function which plays and/or transmits an alarm – a function that would be useful if you are in need and may pass out. Activate the alarm by pressing the [CALL] button and holding it at least two seconds until the alarm sounds. MENU option 32 is used to configure this option.

So, You Bought a Baofeng Radio – Now what?

Note if the radio Alarm option (Menu option 32) is set to TONE or CODE, the alarm sound WILL be broadcast on the frequency you are on – Including across a repeater system. Setting it to SITE prevents it from transmitting and it is sounded over the radio speaker only.

MONITOR Key (6)

Pressing this button momentarily activates the flashlight. Pressing it again changes to a flashing light. Pressing it again turns off the light. Pressing and holding this button turns off squelch allowing weak radio signals to come through.

VFO/MR Key (10)

Not present on the UV-82 series radio, only on the BF-F8HP and other UV-5R Family style of radios. This changes the radio from FREQUENCY mode where you key in your frequency, to Channel mode where you use one of 128 channels that can be programmed.

To perform this function on the UV-82 series of radios, turn the radio off, hold down the [MENU] button, and then turn the radio back on.

A/B Select (12)

This button changes the active channel between A and B for broadcasting. If you are going to save a channel you are listening to you MUST BE in Frequency mode.

Numeric Keypad (7)

The numeric keypad can be used to key in frequencies, jump between menu options, and also used to send DTMF codes through the radio to perform remote functions such as remote repeater control. The layout of the Numeric Keypad will be different between the BF-F8HP/UV-5R Style of radios and the UV-82 style, but all the

So, You Bought a Baofeng Radio – Now what?

keys on the keypad itself are on both radios. (The UV-82 has no VF/MR Button however)

The Pound [# ▯O] key

On the Baofeng radios, this key also acts as a power shift button. Pressing the key on the radio will alter the transmit power between L, M and H. If you change the power to a channel you are using, the power change only remains for that communication session until the radio is powered off and back on – It does not save to the channel's memory. Also switching to another channel or mode (Frequency or FM Radio) will also reset the power to the channel's programmed setting.

Holding this key down for 2 seconds will also activate a keypad lock for all keys except for the side keys. This is handy to keep the radio from being changed by bumps. Unlock the radio by again holding down the [# ▯O] for two seconds.

The [*] Key

If you are talking on a repeater then this key will allow you to swap your frequencies for Transmit and Receive for that channel. This can be useful when troubleshooting communications on a repeater that you think you may be having problems with.

The [0] (Zero) Key

This key on both style radios will display the battery voltage if held down for 2 seconds.

So, You Bought a Baofeng Radio – Now what?

The [BAND] Key
This key is only on some UV-5R Style radios and performs a rotation between bands (UHF/70cm, VHF/2-Meter, and 220/1.25-Meter) when the radio is in FREQUENCY mode only.

The [MENU] Key
This little key is the doorway into the menu of 40 or 41 functions depending on the model of your radio. This area goes so deep that we have reserved the next chapter specifically to the MENU.

The [Exit/AB] key
This key exits you out of the menu functions if you are in the menu.

On the UV-82 series of radios this key, when not in MENU mode, switches you between channels A and B.

Note – To save frequencies to a channel memory you MUST be on Channel A.

When listening to the FM Radio on the Baofeng, the [Exit/AB] key will switch between 65-75 MHz frequency range and 76-108 MHz range. In the United States, FM Radio is strictly on the upper range of frequencies.

The [PTT] Key or Keys (Push-To-Talk)
On the original UV-5R and continuing in the BF-F8HP style of radios, there is a single PTT key on the left side of the radio. The pushing of this key activates the transmitter to broadcast what you say through the Microphone. For the older style of radio, the broadcast goes out through either the A or B channel, depending upon which is active and set through the A/B button.

So, You Bought a Baofeng Radio – Now what?

On the newer UV-82 series of radios, these radios have a PTT Rocker key/Dual key which allows you to effortlessly talk across both channels. Pressing the PTT Rocker upward transmits on the A channel while pressing the PTT down transmits on the B channel. The PTT dual function can be disabled so that pressing it either up or down transmits only on the active channel. This is done through the Baofeng software that comes with the radio (Not through CHIRP).

Where you buy your radio may matter
Baofeng radios are easily available for as little as about $20. But when looking on Amazon you will see many for much more. So, what's up?

You need to understand that there are several dealers who sell the radios online. For instance – Radioddity sells the UV-5R Plus (5-Watt) for as little as $25.00, where Baofeng Tech sells the BF-F8HP (8-watt) radio for over $60. Both resellers sell slightly different models. The 8-watt GT-3TP from Radioddity is still only $35. So again, why the difference?

The difference is service and support. If you are familiar with the radio and realize it is often a throw-away product, you can save money with the cheaper source. But if you need warranty service you may have to send the radio back to China at a shipping cost that exceeds the cost of the radio. Purchasing from Baofeng Tech allows you to return the radio to South Dakota for a quicker turnaround for much less cost.

Baofeng Tech sells several models that THEY have obtained modified firmware for and have had these radios tested and approved by the FCC for specific use. These include the GMRS-V1,

So, You Bought a Baofeng Radio – Now what?

MURS-V1 and the UV-82C which have the proper FCC Part 90 or 95 certifications. If you standardize your business with the Commercial radios from them then this will be the best cost-effective source.

This description of where to buy is NOT a preference for a vendor, however a notation. If you consider the radio a throw-away product you can save money with a cheaper source. But if you need support and service consider the source.

The BAOFENG LCD Display
(Most Models)

LCD Icon Summary:

Icon	Description	Icon	Description
188	Memory Channel	R	Reverse Function enabled
25, 75	Least Significant Modifiers	N	Narrowband Enabled
CT	CTCSS enabled	🔋	Battery level indicator
DCS	DCS Enabled	🔒	Keypad locked enabled
+,-	Frequency shift direction if enabled in VHO	H, M, L	Transmit power level (High, Medium, Low)
+-	Frequency Shift Direction if enabled in MR	▲▼	Indicated active band or channel
S	Dual Watch Enabled	📶	Squelch Open/Close indicator
VOX	VOX Enabled	VOX	VOX Mode turned on

So, You Bought a Baofeng Radio – Now what?

Notes on setting up the Baofeng radios for Amateur Radio use

In contrast with Commercial radio operators, who often need very specific requirements to be compatible with a very specific radio implementation, Amateur radio operators tend to need the broadest possible settings in order to be compatible with as many systems as possible. This basically implies turning all the fancy features that you typically might need for a commercial setup off.

In a typical Amateur radio setup, the following settings would be recommended:

(1) Set bandwidth to Wide (menu item 5).

(2) Turn DCS and CTCSS off (menu items 10 through 13).

(3) Turn ANI, DTMFST, S-CODE, PTT-ID off and PTT-LT to 0ms (menu items 15 through 17 and 19 through 20).

(4) Turn off Squelch Tail Elimination (STE) features (menu items 35 through 37).

(5) Turn roger beep (ROGER) off (menu item 39).

So, You Bought a Baofeng Radio – Now what?

So, You Bought a Baofeng Radio – Now what?

Chapter 3: Navigating the Radio MENU system

One of the more difficult aspects of getting comfortable with a new radio is understanding the menu system. For me the Baofeng seemed overwhelming when I started using it. So much so that I quickly moved to using CHIRP which did make using the radio and setting it up much easier. Later when I picked up my Yaesu radio and by Baofeng Tech DMR Radio I had to dig deeper into the programming by hand method.

My intent here is to cover the menu system used in most Baofeng radios and review a little deeper what each menu option does and review the values available for each option.

0. **Squelch – Squelch level (Default 5)**
   The Squelch level mutes the radio speaker if there is no strong signal. Setting values range from (0) – Off, up through (9). The higher the level of squelch the stronger the signal will have to be to be heard.

   The squelch on this radio is a DIGITAL squelch with settings ranging from (0) (Squelch turned off) to a maximum of (9) which only allows the strongest of signals to come through. Setting the radio to 0 is the same as pressing and holding [Side key 2] on the UV-82 radio, or the [Monitor] key on the BF-82HP (UV-5R series) which turns squelch off while being held down.

   Note that the [CALL] button is not functional when [MENU] (0) = 0

1. **Step – Frequency Step (Default 2.5K)**
   This step value is the frequency step size in VFO/Frequency mode when the radio is scanning or pressing the [UP] and [DN] keys.

So, You Bought a Baofeng Radio – Now what?

Values for this setting range from the lowest setting of 2.5K up to 50K, but not necessarily in even increments. Adjusting this value from the default of 2.5K up will increase the step size during scanning and speed up the frequency scan, however you will also skip frequencies in the scan that could have signals coming across.

2. **TXP – Transmitter Power**
   Allows to select between LOW, MID and HIGH transmitter power.

   This value will vary depending on the specific radio. For the smaller 4-watt radios, the options will be 1w/4w with setting options of LOW and HIGH only. On the higher watt radios such as the UV-82HP and the BF-F8HP radios, the options are LOW, MID and HIGH at wattages of 1w/4w/7w or 8w. The difference on the High side is because the radio transmits at 7w for UHF, but 8w for VHF frequencies.

   Note that the power level can be toggled in MR/Channel mode by pressing the [# ᵣO] key and may require Menu 7 to be set to (0).

3. **SAVE – Battery Save** (Default of 3)
   This setting allows the ratio of sleep cycles to wake cycles be adjusted from 1:2, 2:1, 3:1 or 4:1. The higher value extends the battery life the most. When turned on, there is possibility of missing a portion of the received transmission when the monitored frequency activates.

   The life of the Baofeng batteries will vary depending upon the size of the battery and the use of the radio. Another factor is the setting of menu option 7, TDR ON. When the radio is in dual-watch mode (Option 7 turned on) then the receiver is alternating between both channels for monitoring. The battery save mode kicks in after the

So, You Bought a Baofeng Radio – Now what?

radio does not receive for about 9 seconds, and or no keyboard activity.

- SAVE=OFF: Power saving disabled.
- SAVE=1: RX on ~0.2 S, RX off ~0.13 S.
- SAVE=2: RX on ~0.2 S, RX off ~0.23 S.
- SAVE=3: RX on ~0.2 S, RX off ~0.33 S.
- SAVE=4: RX on ~0.2 S, RX off ~0.43 S.

4. **VOX – Voice Operated Transmit** (Default is OFF)
When turned on the microphone will activate automatically based on set sensitivity.

The VOX mode is turned off by default. With the settings of 1 through 10, 1 represents the most sensitive value where 10 represents the setting require the strongest voice.

Note that when VOX is not set to OFF, 'VOX' is indicated in the status display.

5. **WN – Wideband / Narrowband** (Default is Wideband for Ham radio)
Controls the bandwidth setting of 12.5 kHz or 25 kHz.

Ham radio uses the older standard of 25 kHz. Many radio systems including Commercial radio are moving towards narrowband of 12.5 kHz. Though the Baofeng radios are not FCC Approved for use outside of HAM radio, these radios do have this capability. Baofeng Tech's UV-82C Does have the FCC Part 90 Approval allowing it to be used for some commercial uses.

When this is set to NARR, an 'N' is indicated in the status display.

So, You Bought a Baofeng Radio – Now what?

6. **ABR – Display Illumination Time** (Default is 10)
   Sets the timeout for the LCD Backlit screen in seconds. The default is 10 seconds. Reducing the number of seconds help to save battery life.

7. **TDR – Dual Watch, Dual Reception** (Default is ON)
   Sets the dual watch to monitor channel A and Channel B at the same time. The Baofeng can monitor two channels by alternating between each channel. The radio has only one receiver, but switches between the two for monitoring. If a communication is received the radio will broadcast the signal being received on either channel. While receiving on that channel the radio will remain on that channel, then will return to monitoring after the transmission closes.

   By turning the dual reception to off, the radio will not alternate reception between the two channels and will remain on the active channel only.

   Note: TDR should be turned off when programming the radio manually.

8. **BEEP** (Default is ON)
   Turns on or off the Keypress beep. When on, the radio will beep when the keys are pressed. To silence this, turn the BEEP setting to OFF.

9. **TOT - Transmission Time-out-Timer** (Default is 60)
   This setting turns off the transmitter after a set time as a way to prevent long transmissions and save battery life. This also helps to prevent transmitting if your PTT button becomes stuck.

   Values are in 15 second increments starting with 15 seconds (Setting

So, You Bought a Baofeng Radio – Now what?

of '0') and up to 600 seconds (Setting of '39'). Be aware if you are talking regularly for long periods, such as on a radio NET you may need to extend this out. Many repeaters are set with a timeout to prevent over-use and 2 minutes is common on some repeaters. If your radio is set to 60 seconds, your communication will cut at 60 seconds possibly before you are finished.

The red TX LED on some radios may flash for 10 seconds before the end of the timeout.

10. **R-DCS Receiver DCS** (Default is OFF)
    This option mutes the speaker of the transceiver unless a specific low-level digital signal is received. Without the signal coming though you will not hear anything. The values are provided in Appendix K: DCS Codes. This can be used with other radios along with Menu setting 12 (T-DCS – Transmitter DCS). If another radio has T-DCS signal set to the same you are set to receive, then their transmission will come through. Otherwise your radio will ignore all other transmissions.

    Setting R-DCS sets Menu 11 to OFF.
    Recommend setting this to OFF.

    It is important to know that this is a method to filter out unwanted transmissions – But anyone with a radio set to R-DCS OFF will hear all transmissions. THIS IS NOT A METHOD OF PRIVACY – Only a method of filtering.

11. **R-CTCS - Receiver CTCSS** (Default is OFF)
    This option mutes the speaker of the radio if there is not a specific and continuous sub-audible signal coming through. The station you are listening to MUST transmit this signal for you to hear them.

So, You Bought a Baofeng Radio – Now what?

This is very similar to Menu item 10 (R-DCS Receiver DCS) but uses a different set of tone types. This can be used with Menu Setting 13 (T-DTCS) for filtering. Appendix J: CTCSS Squelch Codes contains the list of codes that can be used.

Recommend setting this to OFF except for repeater use.

Notes for Menu Item #11:
- When R-CTCS is not set to OFF, CT is indicated to the left of the upper channel display.
- Setting R-CTS sets Menu 10 to 'OFF'

It is important to know that this is a method to filter out unwanted transmissions – But anyone with a radio set to R-DCS OFF will hear all transmissions. THIS IS NOT A METHOD OF PRIVACY – Only a method of filtering.

12. **T-DCS - Transmitter DCS** (Default is Off)
This option allows the transmission of a low-level signal if turned n that will unlock the squelch of a remote radio or repeater.

If your radio is set to use T-DCS, then a receiving radio can also be set to the same T-DCS Code so that it filters out other broadcasts. The values are provided in Appendix K: DCS Codes. This can be used with other radios along with Menu setting 10 (R-DCS – Receiver DCS). If another radio has R-DCS signal set to the same you are using (Or set to OFF), then your transmission will come through.

It is important to know that this is a method to filter out unwanted transmissions – But anyone with a radio set to R-DCS OFF will hear all transmissions. THIS IS NOT A METHOD OF PRIVACY – Only a method of filtering. Recommend setting this to OFF.

So, You Bought a Baofeng Radio – Now what?

13. **T-CTCS - Transmitter CTCSS** (Default is Off)
    This option transmits a continuous subaudible signal that will unlock the squelch of a distant receiver which is usually a repeater. Appendix J: CTCSS Squelch Codes contains the list of codes that can be used.

    Repeaters often require a CTCSS tone to be transmitted from the radio transmitter to trigger the repeating function. If the transmitter does not send the proper CTCSS tone the repeater will ignore the transmission.
    Recommend setting this to OFF except for repeater use.

14. **VOICE – Voice Prompt** (Default is English)
    This controls the voice prompt for when a button is pressed. Options are ENGLISH, CHINESE or OFF.

    This is the voice that occurs when you press the [MENU] button or press buttons where the radio provides a verbal response.

15. **ANI-AD Automatic Number ID** (This MUST be set through software – Not set through the menu)
    This displays the ANI-ID code set by software - Cannot be changed through the Menu. This value can be modified through CHIRP after downloading the radio configuration, then SETTINGS → DTMF Settings.
    The ANI-AD is also known as the PTT-ID and is a code transmitted along with your voice transmission which identifies your radio type. This is commonly used in selective calling/signaling systems usually used in commercial and public safety systems.

So, You Bought a Baofeng Radio – Now what?

16. **DTMFST – DTMF Side Tone of transmit code**
    (Default is DT+ANI)
    This option is used to determine when DTMF Side tones can be heard from the radio speaker.

    Values range from (0) to (3). DTMF Tones are keypad tones for the numbers 0-9, letters A, B, C &D, the Asterisk (*) key and the pound (#) key. These tones and the keypad can be used to send audible tones in a transmission for remote activation or control of a remote station that listens for the tones.

    Values for this option are:
    - [0]: OFF No DTMF Side Tones are heard DT-ST;
    - [1]: Side Tones heard only from manually keyed DTMF codes ANI-ST;
    - [2]: Side Tones heard only from automatically keyed DTMF codes DT+ANI;
    - [3]: All DTMF Side Tones are heard (Default)

17. **S-Code – Signal Code** (Default is 1)
    The S-CODE options selects 1 of 15 DTMF Codes which are programmed with software, and up to 5 digits. These codes can be modified in CHIRP.

    Values are:

Option 1, Value '0'	Option 6, Value '5'	Option 11, Value '10'
Option 2, Value '1'	Option 7, Value '6'	Option 12, Value '11'
Option 3, Value '2'	Option 8, Value '7'	Option 13, Value '12'
Option 4, Value '3'	Option 9, Value '8'	Option 14, Value '13'
Option 5, Value '4'	Option 10, Value '9'	Option 15, Value '14'

    But what is a Signal code? The signal code is a DTMF code that can be associated with a channel on the radio and broadcast with your transmission. This code can be used by a radio dispatcher (If the radio is used with a dispatch system) to identify the radio and the person transmitting. This Signal Code is a PTT-ID Signal code and can be sent at BOT (Beginning of Transmission), EOT (End of Transmission), or both BOT & EOT.

18. **SC-REV - Scanner Resume Method** (Default is "TO")
This sets the method for resuming scanning - Options are "TO", "CO" or "SE".

What this is used for Is to tell the scanner how to react when running in Scan mode. While in Scan mode the radio can scan either frequencies (In frequency mode) or Channels (In Channel mode). When it comes upon a frequency or channel broadcasting it will stop to let you listen. Then it will do one of the following:

- TO Mode (Time Operation) – The scanning will start back up after a certain amount of time lapses.
- CO Mode (Carrier Operation) – The scanning will start back after the active signal stops.
- SE Mode (Search Operation) – Scanning will NOT resume but will remain on the channel it finds here.

19. **PTT-ID - When to send the PTT-ID (Default** is OFF)
This determines when to send PTT-ID codes sent during either the beginning or end of transmission.

There are Four options for this:
- [0] = OFF, no PTT-ID is sent;
- [1] = BOT = Beginning of Transmission;
- [2] = EOT = End of Transmission;

So, You Bought a Baofeng Radio – Now what?

- [3] = Both = Both BOT & EOT

This ties in with other Menu options to determine exactly what PTT-ID is sent. For simplex communication or communication with a Repeater this is not necessary. It is more likely to be used when working in a radio dispatch system to identify which radio is transmitting. The PTT-ID Code that is set is established in MENU item 17 as to which of the 15 possible codes should be sent, and the actual codes are set via Software programming such as with CHIRP. This is recommended to be left off unless specifically needed.

20. **PTT-LT - Signal code sending delay** (Default is 5 milliseconds).
    This menu option sets the PTT-ID LAG TIME in Milliseconds. Setting values range from 0 to 50. Menu 19 must be enabled for this to work.

21. **MDF-A - Channel Mode A Display**
    (Default is FREQUENCY)
    This sets the display mode of channel A. Display modes can be FREQ for Frequency; NAME for channel name (Set through Software only), or CH for Channel Number.
    When programming your radio with channels, if you program channel "A" for NAME and channel "B" for Frequency, you have the ability to view the NAME reference to the channel on "A" while also being able to view the frequency of the channel on "B".

22. **MDF-B - Channel Mode B Display** (Default is FREQUENCY)
    This sets the display mode of channel B. Display modes can be FREQ for Frequency; NAME for channel name (Set through Software only), or CH for Channel Number.

So, You Bought a Baofeng Radio – Now what?

When programming your radio with channels, if you program channel "A" for NAME and channel "B" for Frequency, you have the ability to view the NAME reference to the channel on "A" while also being able to view the frequency of the channel on "B".

23. **BCL - Busy Channel Lock-out** (Default is OFF)
    This option will disable the [PTT] button if that channel is already in use. The radio will beep and will not transmit if the channel is in use. Values for this are either ON or OFF.

    This feature is useful for preventing "Doubles" or two individuals keying and talking on top of each other.

24. **AUTOLK – Automatic Keypad Lock** (Default is OFF)
    Turning this ON locks the keypad if not used in 8 seconds. To unlock, press the [# ꭲO] key for 2 seconds. Options for this are ON or OFF.

    This is useful to prevent you from accidentally rubbing against or pressing the radio keys which could change the operation of the radio. This only affects the buttons on the FRONT of the radio and not the PTT or CALL buttons.

25. **SFT-D - Frequency Shift Direction** (Default is OFF).
    Selects the Repeater Frequency Offset Direction (-) or (+). Values for this are:
    - [0]: OFF - TX = RX (simplex);
    - [1]: TX will be shifted higher in frequency than RX (+);
    - [2]: TX will be shifted lower in frequency than RX (-);

    When this is set to (+) or (-) the LED Display will indicate which shift is occurring with a + or – symbol. This is used with Menu option (26) to access repeaters. This is NOT needed when storing repeater frequencies.

26. **OFFSET - Frequency shift amount** (Default is 0)
    This sets the offset amount between TX and RX frequencies. Values for this range from 0 to 69.990 kHz in 10 kHz steps. For VHF the offset in the US is set to 00.600 MHz (600 kHz) for VHF frequencies, and for UHF Frequencies in the US 5 MHz is the most common setting.

    This is used in conjunction with Menu option 25 when setting up repeaters.

27. **MEM-CH - Store a Memory Channel**
    Use this option to create a new channel or modify an existing one for access in Channel mode.

    To store a new channel, you will need to set the radio first to FREQUENCY mode so that you can enter the characteristics. You will need to add the frequency, offset, CTCSS Tones, and any other custom settings needed. If you are setting up a SIMPLEX channel there will be few settings. Repeater frequencies will require more settings.

28. **DEL-CH - Delete a memory channel**
    This menu option is used to delete the information programmed into an existing channel for reprogramming or leaving empty.

29. **WT-LED - Display backlight color, Standby** (Default is Purple)
    Use this menu option to set the display backlight color for standby.
    - [0]: OFF;
    - [2] = ORANGE;
    - [1] = BLUE;
    - [3] = PURPLE;

So, You Bought a Baofeng Radio – Now what?

30. **RX-LED - Display backlight color, receive** (Default is Blue)
    Use this menu option to set the display backlight color for standby. Options are:
    - [0]: OFF;
    - [1] = BLUE;
    - [2] = ORANGE;
    - [3] = PURPLE;

31. **TX-LED - Display backlight color, transmit** (Default is Orange)
    Use this menu option to set the display backlight color for Transmit. Options are:
    - [0] = OFF;
    - [1] = BLUE;
    - [2] = ORANGE;
    - [3] = PURPLE;

32. **AL-MOD - Alarm Mode** (Default is TONE)
    You have three options here which are: SITE=This sounds the alarm through the radio speaker only; TONE=This sends a cycling tone transmitting over the air and Speaker; or CODE=Transmits '119' followed by the ANI code over the air (Reverse of 911).

    Using the alarm on the radio will sound an audible alarm over the speaker, or Transmitted over the air and speaker. This can be used if you have an emergency and feel you will not be able to call out for help allowing others to hear the radio, and if you are with a group on the radio then others will know someone is in need of help.

    It is very important to remember that when set to TONE the radio WILL BROADCAST the alert which is a unique sound to Baofengs over the air. Prepare to take grief by other hams if you do this on a repeater frequency. Most hams will recommend turning this off

So, You Bought a Baofeng Radio – Now what?

which is not possible, so consider changing it to SITE which will not transmit.

33. **BAND – Band Selection** (Default is VHF)
    This menu item allows you to select the default band when in VHF/Frequency mode. When in Channel mode the last channel you used will be your default when turning on the radio.
    You have only two options for this – VHF or UHF.

34. **TDR-AB - Transmit selection while in Dual Watch mode** (Default is OFF)
    Use this to set the priority to the selected display (A or B) when the signal in the opposing display stops. Since the Baofeng has two channels that it can monitor, this setting, along with setting Menu item [7] to ON, tells the radio which channel to return to after monitoring and activating for a channel.

35. **STE - Squelch Tail Elimination** (Default is ON)
    Use this to eliminate the squelch tail noise that occurs from Baofeng handheld radios that are working in simplex mode (Radio-to-Radio without a repeater). Values are ON or OFF.

    When enabled and T-DCS is set to OFF the radio sends a 55 Hz tone for about 1/4 second when the PTT key is released. When enabled and T-DCS is not set to OFF the radio sends a 134.4 Hz tone for about 1/4 second when the PTT key is released. Set to OFF before communicating through a repeater.
    It is recommended to set this to OFF.

36. **RP-STE - Squelch Tail Elimination** (Default set to 5)
    Use this to eliminate the squelch tail noise when operating through a repeater. Values for this range from 1 thru 10. Requires use of a

repeater utilizing this feature. Used with menu [37]. Recommended setting is OFF.

37. **RPT-RL - Delay the squelch tail of repeater** (Default is OFF)
    This setting delays the tail tone of the repeater in milliseconds after the [PTT] is keyed. Values range from [0] to [10]. Recommended setting is OFF.

38. **PONMSG – Power-On Message**
    This controls how the display acts when the transceiver is turned on. Values are FULL or MSG. The Power-on message must be edited through software

    FULL – Performs an LCD screen test at power-on.
    MSG – Displays a 2-line power-on message.

39. **ROGER - Roger Beep** (Default is OFF)
    This sends an End-of-Transmission tone to indicate to other stations that the transmission message has ended. Values are OFF or ON. Recommended setting is OFF.

40. **RESET - Restore defaults** (Default is ALL)
    Performs a factory reset on the radio. The options are:
    - VFO: Resets all of the menus to factory default and sets the [A] and [B] VFO Frequencies to firmware default.
    - ALL: Resets all of the menus to factory defaults. This does reset all channels and programs channel (0) to 136.125 MHz and channel 127 to 470.625 MHz.

41. **R-TONE – Repeater Tone** (UV-82 Only)
    The R-TONE is short for Repeater Tone and is used to activate repeaters that require a specific audible tone to be transmitted for

So, You Bought a Baofeng Radio – Now what?

access. Transmit by pressing the [F] side key while the [PTT] button is also pressed. UV-82x series only. Values for this are:
- [0] 1000 HZ;
- [1] 1450 HZ;
- [2] 1750 HZ;
- [3] 2100 HZ;

Menu items affecting the Battery or Transmit power:

Menu #	Title/Description
2	TXP – Transmit Power This sets the actual Transmit power. Set to LOW power to conserve battery during Transmit.
3	SAVE – Battery Saver Setting this value to a higher value will reduce the frequencies that the radio polls frequencies for signals. This can result in receiving partial transmissions.
6	ABR – Display Illumination Time Reducing this from 10 seconds will reduce help extend battery life.
7	TDR – Dual Watch, Dual Reception Turning this setting to OFF will stop the radio from polling both frequencies for channel A & channel B. Stopping the polling will conserve battery life.
9	TOT – Transmission Time-out-Timer Reducing the max transmission time will reduce long transmissions and help extend battery life.

So, You Bought a Baofeng Radio – Now what?

Menu items related to Repeaters (In the US)

Menu #	Title/Description
11	R-CTCS - Receiver CTCSS This option is used for Repeater tone signals. This value varies between repeaters and not needed for Simplex.
13	T-CTCS - Transmitter CTCSS This option is used for Repeater tone signals. This value varies between repeaters and not needed for Simplex.
25	SFT-D – Frequency Shift Direction This establishes the frequency shift direction (+ or -) when using a repeater
26	Offset – Frequency shift amount Sets the amount of the frequency offset (Usually 600 kHz for VHF or .6 MHz, and 5.0 MHz for UHF)
36	RP-STE – Squelch Tail Elimination Eliminates squelch tail noise when operating with a repeater. For HAM leave this OFF.
41	R-TONE – Repeater Tone (UV-82 series only) A Repeater tone used to activate repeaters that require a specific audible tone for access.

So, You Bought a Baofeng Radio – Now what?

Menu items recommended to be turned off for HAM Radio use:

Menu #	Title/Description
10	R-DCS - Receiver DCS This option is used for filtering signals along with menu option 12. Setting this may prevent you from hearing some transmissions.
12	T-DCS – Transmitter DCS This option is used for Repeater tone signals. This value varies between repeaters and not needed for Simplex, so turning it off by default is recommended.
19	PTT-ID (Off or When to send) Not needed for HAM Use. This would be used with a dispatch radio system.
32	AL-MOD – Alarm Mode Because this will transmit the alarm, and will transmit over REPEATERS, we recommend turning this off so you don't accidentally send an alarm code across a repeater network.
35	STE – Squelch Tail Elimination Recommend setting to OFF before communicating with a repeater to avoid problems.
36	RP-STE – Squelch Tail Elimination (With a repeater) Recommend leaving turned off for HAM radio use
37	RPT-RL – Delay the squelch tail of repeater Recommend leaving turned off for HAM radio use
39	ROGER – Roger Beep Recommend leaving turned off for HAM radio use as it becomes annoying to other operators

Chapter 4: REAL Radio Capabilities

Let's talk about the real capabilities of the radio. Not what many people THINK it can do, but rather what you will really get out of the radio in a real, practical and everyday use. These handheld radios can be used to communicate using two basic methods – SIMPLEX mode in which you are talking from radio Operator-to-Operator, and in a REPEATER mode in which you are talking with the radio Operator-to-Repeater-to-Operator.

Operator-to-Operator
In this mode, each operator is holding the radio and talking to each other. The distance you can get between the operators depends on the following factors:

- Antenna Height from the ground
- Power of the radio
- Interference and obstacles between the two operators

These radios work in frequencies that are basically "Line-of-Sight", meaning signals generally do not bounce in the atmosphere and do not bounce well off of objects. Due to the curvature of the earth, the higher each antenna is from the ground then the farther the distance between two operators. There is an actual formula which can be used to determine what your line-of-sight range will be based on height of the antenna from the ground. This formula for the radio horizon is:

$$\text{Horizon (In Km)} = 4.12 \times \sqrt{\text{Height (In Meters)}}$$

Using this formula, let's assume we have two radio operators where they are holding their radios exactly 2 meters in height (About 6'6"), then we take the square root of 2 which comes out to 1.4142.

So, You Bought a Baofeng Radio – Now what?

We now multiply 1.4142 by 4.12 and we get a distance of 5.82 km which is equal to about 3.6 miles.

Since we have two operators, each operator has a radio horizon of 3.6 miles, so we have a distance of 7.2 miles between the two. This is in an optimum situation – nothing between them. No trees, buildings, leaves, or anything. Flat earth – such as on the ocean, or in a completely flat desert.

Now let's add some of those obstacles. Buildings, cars, trees, leaves, hills, and we severely cut down the range of our two operators to usually two to five miles, and in most cases towards the lower numbers. But – Remember – Height of the antenna is critical. More important than power of the radio.

Operator-to-Repeater-to-Operator

In this second method, there is a repeater somewhere high up that we can reach much more easily. Around me I frequently hit a repeater that is 850 feet on a tower. Let's see what my radio horizon is with this repeater:

First convert to Meters: 850 ft = 259 meters

So, my radio horizon is: $4.12 \times \sqrt{259}$;
or... 4.12×16.09

Making my Radio horizon = **66.29km**, or **41.19** miles. (1 KM = .62 Miles)

And as the commercial host says – But wait, there's more…

We must add the 3.6 miles taking into account my standing height of 2 meters, so our total radio horizon to the repeater is 44.79 miles or

So, You Bought a Baofeng Radio – Now what?

rounded up to 45 miles. Now if my buddy is 45 miles on the other side of the same repeater, then we can talk about 90 miles apart.

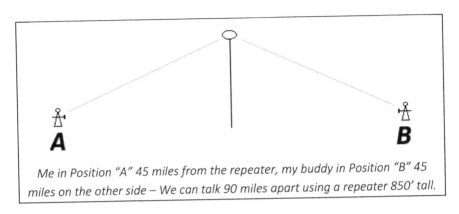

Me in Position "A" 45 miles from the repeater, my buddy in Position "B" 45 miles on the other side – We can talk 90 miles apart using a repeater 850' tall.

But wait – there's more...

The repeater that I connect to is a networked repeater. It is part of a repeater network across the state, so when I talk and hit the repeater I am instantly rebroadcast as far as Panama City Florida, or as far south as Miami – A full 12-hour driving distance between the two cities. So here you can see there is great distance capabilities with your simple Baofeng handheld radio.

In a Repeater network as represented by this image of Florida, all of the dots represent a repeater location interconnected. Talk on one repeater and you can be heard by all on all repeaters.

Now as for reaching that repeater tower – I said the Antenna is the most important. Both quality of your antenna and the height of the antenna

So, You Bought a Baofeng Radio – Now what?

play a huge role in getting a signal to the repeater. And of course, the closer you are to a repeater or to another person you are talking to the better the signal. But this is also where signal strength is very important. If I reach the repeater or other operator with a weak signal, they may not be able clearly hear me or I may not have a strong enough signal to trigger the repeater to rebroadcast my signal. In this case, a higher power radio will make a difference. Now my handheld may be good at picking up signal from 45 miles away, but I may have a problem pushing a signal that far with a handheld. But – I can certainly push a signal ten, or maybe twenty, or maybe more miles away. This is where the interference thing comes in.

If I want to push a strong signal, from my Baofeng I do have some recommendations. Baofeng Tech does sell a signal amplifier. These are unique for the specific band you need to transmit on – one of VHF and a different model for UHF. They connect to your antenna feed out and will boost your signal to 40 watts. This is one way to push extra power. But as for the raw radio itself, your best enhancement you can get is to update your antenna.

The best single update that you can do on a stock Baofeng radio, or any radio for that matter, is to update your antenna. When we talk about antennas later in Chapter six, I will give you some examples of alternative antennas that you can get for your radio. On Baofeng Tech's own website in fact, they market several brands as optional updates to the stock antenna that comes with Baofeng radios. Several of these are models of the popular Nagoya antennas. The short stock antenna that comes with the radio works adequately for very short distances but getting a better antenna for the radio should be one of the top items on your list.

So, You Bought a Baofeng Radio – Now what?

If you will be using the radio in the car, then a car-mounted antenna should also be added to your list. A trunk mount antenna or a good Magnet-mounted antenna will set you back under $40 but will move your reception and transmission point outside of the car and out of the signal shielding that the body of the car will act as. You will find weak signals come in strong, strong signals come in stronger, and signals that you would otherwise not have heard can be heard. The same works for transmitting allowing others to hear you while in your car that otherwise would not have been able to.

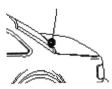

What the Baofeng Radio is <u>NOT – Not Digital</u>

In this book we are focusing on the capabilities of the Baofeng Radios and covering what these radios can actually do. But if your new to learning Ham radio, there will be some things that you will hear and I want to clarify some things on the radios as well as Tech License capabilities to have a clear understanding of what the radio is not.

These Baofeng radios are ANALOG radios – That means they operate strictly using analog radio signals vs. the newer Digital radio systems. Analog signaling is what has been around since radio was invented. This is also going to be the most reliable in an emergency/radios down scenario because Analog has fewer dependencies on technology.

By saying these radios are Analog, we are saying they are NOT digital systems. Baofeng does manufacture a Digital DMR Radio system – It also works with analog as well, and costs close to $100. A Digital radio or DMR radio communicated through a DMR Repeater, usually connected

So, You Bought a Baofeng Radio – Now what?

to an Internet backbone. Additionally, you can connect a DMR radio to a special hotspot that acts as a bridge between your radio and the Internet allowing you to then connect to talk groups, through the Internet, and to other users around the world. These radios can be great – but again, different technology and different capabilities.

Other types of Digital radio include P.25, D-Star, Yaesu System Fusion, NXDN and others. Most municipalities are converting to Digital based radio systems including Police, Federal, FEMA, Fire and others. Digital does have great advantages in use including the ability to get more conversations within the same bandwidth, Time Divided multiplexing, security and encryption (Which is NOT allowed on HAM), and better signal clarity.

As a licensed Technician you do have the ability to play in this area. Note that most digital radio systems are much more costly than your inexpensive Baofeng radio – in fact you could by between ten Baofeng radios (Or more) for what you have spent on your Baofeng radio that you have here. Of all of the digital modes, DMR in the United States has gotten a footing with models being brought to market by many manufacturers including Baofeng. The Baofeng DM-1701 & DM-1801 are two from the manufacturer Baofeng and available in the United States. Baofeng Tech sells under their name the DMR-6X2 radio which is actually an AnyTone AT-D878 radio with updated firmware and capabilities allowing it to work both Digital DMR and Analog modes.

If you decide to venture into Digital there are some good low-cost (Sub $200) radios available, but make sure the radio you get also works Analog. I own the DMR-6X2 as my go-to radio. I purchased this radio as an upgrade to my other BaoFeng radios in part because it does support Analog communication, but it also supports a wide range of features including over 4,000 programmable channels. But along with capabilities

So, You Bought a Baofeng Radio – Now what?

comes complexity making that radio far outside the scope of any further discussion.

IN SUMMARY

So, in summary you need to know that the actual Operator-to-Operator range without a repeater is only going to be between two and five miles for these radios. But by connecting via a repeater you can greatly extend this range which will allow you to connect much farther.

Quality of your antenna is critical and there are several options you have. We will recommend looking at another antenna to increase your radio's performance and have listed three good recommendations to try that will replace the antenna on the radio itself. Connecting to a Car mounted or externally mounted antenna will also go a good way in extending the range of your radio as well as signal performance.

In 2018 when Hurricane Michael tore through North Florida, there was a story of a family literally trapped. They lived on a back-road area where the storm had devastated the trees and area around them. Though they were high and dry, and their home survived the storm, but they were so off the beaten path that they could not drive out of the area or walk out due to age and health reasons. Had they had a small handheld radio they would have easily been able to call for help as the area repeaters did remain operational and they would have had a good 20-mile range on the radio. Instead they were forced to write out "Help" in the yard and wait days for someone to see it overhead, which a family member actually did from aerial photos of the devastation. All survived healthy and well, but this is an example where a radio capability could have prevented disaster for the family.

So, You Bought a Baofeng Radio – Now what?

Chapter 5: Understanding VHF & UHF

Your radio works on two bands – VHF and UHF. Both of these are considered to be "Line-of-Sight" radio signals which means the signals do not reflect like HF Radio signals or even CB Radio signals will, but rather they will generally travel the distance of the radio horizon. We talked about the radio horizon in the last chapter when we talked about realistic radio capabilities. To refresh, the radio horizon is the distance the radio signal will travel based upon height of the antenna and curvature of the earth.

VHF and UHF are the same bands that you use with your Television and Car Stereo to pick up local channels through an antenna. Think about your reception from various radio stations or TV stations around you. The farther you go, the weaker the signal and the more noise you get. As bands go down in frequency, characteristics such as atmosphere skip start to come in to play where we can pick up longer distance signals. Think of listening to AM Radio – Sometimes depending upon the weather and atmosphere you can pick up signals from much farther away. The AM Radio signals are in the lower kHz frequencies – much longer wave forms which can at times travel much farther. Shortwave radio also uses these longer wave forms to travel farther – sometimes thousands or tens of thousands of miles. Ham radio as we know it from Television and Movies is usually focused on HF radio which you can use when you get your General level license – covering much farther distances around the country and around the world.

But your Baofeng – It is limited to VHF and UHF. Ever see police officers out of their cars with their handheld radios? These communicate usually on UHF frequencies and to local area repeaters which connect them to their central dispatchers. Government and Public safety systems use sophisticated digital communications and complex Trunking systems to

So, You Bought a Baofeng Radio – Now what?

allow maximum use for their available frequencies. But – These are still UHF Line of sight, so they still depend on repeater networks for distance. Without these repeaters they would not get the distance they needed to stay in touch.

In fact – Next time that you see a police car, look for the antenna on the car. You will see a short antenna just a few inches long. The shorter the antenna the higher the frequency. Most public safety radios work in the 800Mhz or higher frequencies which is outside of the range your little Baofeng can communicate on. But remember that UHF frequencies range from 300 MHz up to 3,000 MHz or 3GHz.

We get range with these radios through the use of repeaters. The higher the repeater, the better the range. In the example I gave in the prior channel, a local repeater to me here in the Tampa Bay area is 850 feet up. This height allows me about a 45-mile range away from the repeater – That allows me to travel anywhere within my county and still be able to talk to other hams.

So – Which works farther? VHF or UHF?
Good question. VHF will give you a better range than UHF will. Not by much, but some. This is generally outdoors with a slightly better radio horizon distance. Also, VHF works better with a longer antenna. These short stubby antennas you see on radios will not provide a very good range. If your using your Baofeng outdoors on VHF it is much better to use a full-length antenna.

Which works better within a Building?
From within a building, the higher the frequency then the better the penetration. This means the UHF bands will work better at penetrating through the walls of a building. When using the radios within a building, the shorter antenna length for UHF also becomes an advantage. The

So, You Bought a Baofeng Radio – Now what?

higher the frequency the shorter the antenna and walking around with a shorter antenna in a building is always much more convenient. As a footnote on antenna length – Have you ever thought about why there are not many handheld CB Radios? The antenna length on the 27Mhz band is long – 36 feet for a full-length antenna, and 9 feet for a quarter wave. This is a HUGE advantage for UHF Radios.

What about Signal Reflection?

Radio waves in the VHF and UHF ranges to not reflect through the atmosphere nearly as easily as HF radio waves. (They do under some circumstances but not as the norm). These shorter wavelengths just pass through the atmosphere usually which is another reason for shorter ranges. HF radio waves will have a better reflection quality and will reflect through the atmosphere bouncing back down hundreds or thousands of miles away.

Frequency Range

Our Frequency Range is the actual width of the band frequencies that we must use. As you can see from the table below here, we only have 4 MHz of band width between 144 and 148MHz in the VHF band, while in the UHF Band we have 30MHz of range. For this reason, there are more UHF repeaters that can be found than VHF repeaters.

Radio Band	Legal HAM Range
VHF Band (2-Meter)	144 – 148 MHZ
UHF Band (70-CM)	420 – 450 MHZ

So, You Bought a Baofeng Radio – Now what?

Chapter 6: Power and your Radio

When we talk about power of the radio, we are going to cover two separate topics here.

The first being the actual power methods of your Baofeng radio – The "Juice" – The "Gas" – The Electrical power that your radio runs off of. As a handheld radio, your Baofeng radio works off of a Lithium battery. The radio is recharged from a cradle that is powered from the wall. Is this the only way you can power and charge up your radio?

You actually have a couple of options. With the battery that came with the radio, you need the cradle. But the cradle can actually be connected to a USB to Cradle power plug which will allow you to charge off of a USB power source. This includes being able to charge from a battery pack or your car. The USB Power cable can be purchased from Baofeng Tech or off of Amazon. The cable runs about $12 and can be found at this web address: https://baofengtech.com/usb-charger-transformer.

This adapter will work with any of the Baofeng Tech brand radio cradles, and probably most other Baofeng cradles as well though we cannot say for sure. Best to check the vendor you purchased from.

Though this is more convenient than needing a wall outlet, it is still not a clean solution for when you are in the car. For some of the radios, in particular the UV-5r style radios which include the BF-F8HP radios, there are some third-party extended batteries that work with these model radios that have a charging port on them. One such battery is marketed by Mirkit and comes with a USB Charging cable, as well as being a larger 3800-mah battery.

So, You Bought a Baofeng Radio – Now what?

Many Prepping groups have standardized on the UV-5R modeled radios for the standardization of the batteries. By having multiple members using the same radios, it makes spare battery availability between members convenient. And with the low cost of the Baofeng radios many in the prepping community will purchase multiple radios for family members to have "Just in Case".

The Mirkit marketed BF-F8HP Battery will fit radios modeled after the UV-5R AND has a USB Charging cable allowing you to charge without a cradle.

For the UV-82HP radio and similar models, including Baofeng Tech's GMRS-V1, MURS-V1, and UV-82c Commercial radio, there are extended operation batteries also available though I have not come across any with the USB Charging port on the side of the battery.

For emergency planning and readiness, I strongly recommend at the least getting the USB Charging cable that will allow you to easily recharge from a vehicle or USB Battery pack. Most folks who participate in Emergency planning or prepping will purchase additional batteries to have on hand. These batteries can be charged in their cradles without the radios by the way for convenience.

While talking batteries – let's talk about interchangeability between models of ratios. One of the popular qualities of the radios based on the UV-5R Series is the fact that the batteries are interchangeable. Batteries

So, You Bought a Baofeng Radio – Now what?

from a UV-5R fit a BF-F8HP. Extended batteries or normal batteries can be swapped between radio models.

On the UV-82x series this is MOSTLY true. But not always. I have 3 radios which look virtually the same – The UV-82HP, the GT-3WP, and the UV-82WP models. The second two I purchased through smaller resellers early on and have the radios in part because they are "Waterproof". This is not anything I have actually tested, but I purchased the radios because I like to hike, and I wanted a radio to hike with. What I found out is that all three radios have slightly different batteries which do NOT interchange. The GT-3WP and the UV-82HP radios even have the SAME MODEL battery – Yes, the same part number – But they do not interchange.

I still have the radios, but they are stored away in my emergency comms kit and are not used for regular operation. My UV-82HP I do use regularly, and it has what I consider to be a "Standard" battery that interchanges with my GMRS-V1 radio – both of which I purchased from Baofeng Tech located in the US.

And Now, Another kind of Power

When we talk power there is another type of power that we can discuss for the radio – that is the power of the radio transmitter itself and how much power can be put out by the radios. The Baofeng BF-F8HP and the UV-82HP radios are considered "8-watt" radios. The radios will put out 8-watts transmitting for VHF frequencies, and 7-watts out transmitting for UHF frequencies. This is about twice the power of the original UV-5R radio which actually put out 4-watts of power.

Now when you think this isn't much power, consider that your cell phones put out about a half watt of power or even less. This connects you to a cellular tower several miles away. So, while you are working

So, You Bought a Baofeng Radio – Now what?

with only a few watts with your Baofeng, you are already working with several times more power than your Cellular phone has in it.

But what if we need or just want more power? Additional power from your radio will absolutely make a difference. It will give you a clearer and cleaner signal and makes you better able to reach a repeater or just a remote radio operator. It is not uncommon to be hearing a two-way conversation but only one side of it, even though the party you cannot hear may be closer to you. This can be from a lack of power. Though the party may be between you and the person you do hear, if the person you do hear is pushing out more power, he will easily travel his signal better than someone on a weaker radio.

When traveling I almost exclusively use the mobile Ham radio mounted in my truck. This gives me a better antenna use for reception and transmitting through the permanently mounted antenna, and it also gives me a full 50-watts of power out on the radio. But for a handheld you can still get this higher power by connecting to a radio amplifier. Baofeng Tech sells several such amplifiers that will boost your output up to 50 watts. The amplifiers mount in your vehicle, or you could use them at a desk or base location. Then connect your radio through your antenna connection to the radio input. Now when you transmit the output through your antenna feed is amplified by several times for a stronger cleaner output signal. The Amplifier will connect to a vehicle mounted or externally mounted antenna getting your signal outside of the vehicle or from a better transmitting position.

One drawback about these amplifiers – you need a model specific to the band that you are transmitting on. So, if you connect to a specific repeater or a specific set of repeaters that are within one band this can become a nice solution. If though you are connecting to repeaters in both or multiple bands, then this could be problematic. Keep in mind that there is a lesser used 1.25-Meter band and though most Baofeng

So, You Bought a Baofeng Radio – Now what?

radios cannot use that band, the newer Triband radio from Baofeng, the UV-5X3 radio and several of their mobile radios can transmit on this band.

If you have the need to transmit at a higher power on multiple bands, then rather than consider the amplifier route, you should look at adding a mobile radio to your toolbox. Baofeng Tech sells multiple mobile radios in the 25-watt and 50-watt power ranges.
Ultimately more power, to a degree, is good.

And – By the way -

While I say more power is good – more power in the right place is good. Handheld radios are generally limited to 5- to 8-watts in part for safety. As you increase strength of a radio and place that transmitter to your ear or near your head remember you are increasing the amount of RF energy transmitting near your brain. Recently there have been advertisements for a 16-watt Baofeng radio or other branded radios coming in from China. These are not found from your US Distributors and I would be hesitant to use these radios without a hand microphone. I have also heard of folks having problems locating batteries for these newer and less common radios as they are not the same as the older model radios.

One thing to remember about Baofeng – They are a Chinese company making products for all over the world, but many of their products not designed specifically for the US Market.

So, You Bought a Baofeng Radio – Now what?

So, You Bought a Baofeng Radio – Now what?

Chapter 7: Antennas

The Antenna is, in my opinion, the most important upgrade you can get for your radio. After getting the radio and starting to use it you will want to consider replacing that stock antenna with something better. Also – Let's cover using a simple Mag Mount antenna in a variety of ways to get a stronger signal out.

First things First – the SMA Antenna Mount

All Baofeng handheld radios come with a screw-type SMA Antenna mount on the top of the radio. Now this is fine and works generally well, but the first thing I do to all of my radios when I get them is to install a SMA to BNC Adapter on the radios themselves.

The factory antenna mount and antenna using SMA connectors

A SMA to BNC Adapter screws onto the SMA Mount and allows a BNC Antenna to be added to the radio. Convenient for quick swaps of antenna.

So, You Bought a Baofeng Radio – Now what?

The benefit of adding a SMA-to-BNC adapter to the radio is to allow a quick swap of the antenna and not worrying about mis-threading of the antenna onto the radio. Now electrically this is another connection so there is a slight drawback, but I enjoy the convenience of a quick antenna swap such as when I jump into my less-frequently used car that has a Mag-Mount antenna on it.

Consider if you want to do this to your radio because if you do, you will then want to make sure any new antennas you purchase have a BNC connector to the radio rather than a SMA.

Nagoya Antenna

There are multiple antennas that you can get and put onto your radio in replacement of the stock antenna that comes with the Baofeng which is at best adequate and at worst poor. For less than $20 you can get a Nagoya NA-771 which is a 15.6-inch whip antenna that will operate better for both VHF and UHF with your Baofeng. The Nagoya can be ordered from Amazon and is also available through all Ham radio stores, including Baofeng Tech under their Accessories area. (https://baofengtech.com/accessories)

Nagoya has several good models of their antennas with varying lengths. The NA-771 is just one of several to consider.

So, You Bought a Baofeng Radio – Now what?

Signal Stick by Signal Stuff

Another good antenna you can get is from a company called Signal Stuff and is a $20 antenna. These antennas 19" flexible steel whip antennas which can actually loop on themselves, so they are not sticking up. I have multiple of these myself and I really like them – They are extremely durable as well as being super flexible. Coiling the whip keeps the antenna from sticking up too far, and the coil of the antenna is almost like a trade-mark – Common for users of these antennas. (https://www.SignalStuff.com).

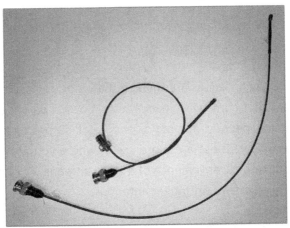

Three Signal Sticks – SMA, BNC and a third coiled. Coiling is done to make the antenna more convenient on your radio when shorter is better.

Abbree Folding Whip

A third good antenna which started becoming popular in 2019 is the Abbree which is a Military-style folding "Tactical" antenna. These are made of a metal ribbon, similar to a tape measure and are able to be folded. The antenna is insulated, and these come in lengths of up to 42.5 inches. Some YouTubers have performed range tests on the antennas to find they actually outperform many other popular antennas. I have one

So, You Bought a Baofeng Radio – Now what?

myself in my WinLink kit that I use a Baofeng radio for. These can be purchased on Amazon also and are under $15.00.

The Abbree folding Military-style whip antenna, 42.5"

"Stubby" style Antennas

These little stubby antennas are marketed as good replacements for the stock antennas that come with many radios. I own a pair of these for my radios (Only one needed per radio) and I get surprising performance from it for receiving. These are more useful where you are not trying to push range but rather convenience of a short antenna. For better range I will always rely on one of my longer antennas.

Diamond SRH805S SMA-F Female Dual Band Antenna and Nagoya the NA-810 2.5" Mini-Whip Antennas

So, You Bought a Baofeng Radio – Now what?

Car-mounted and Mag Mount Antennas

Another type of antenna you can add to your resources is a simple car mounted or a magnet mount antenna. Used with your car, these antennas will allow you to get your transmitting signal outside of the car or truck for greater range. Likewise, you will have greater reception with one of these antennas if the metal of the car is removed as an obstacle from receptions. In fact – Once properly attached, the antenna will be free of the body of the car to get in the way, and the metal of the car will act as a ground-plane for the antenna providing a better signal. The trick is though getting the signal to transmit out from within the car.

A Mag-Mount Antenna is a simple antenna solution

In addition to use in a car, you can also use one of these types of antennas elsewhere to enhance your transmission/reception capabilities of the radio. Some Hams will take a magnet mount antenna and attach it in the middle of a Pizza pan which then acts to radiate the antenna signal out. Or place one of these antennas to a pole or in a tree for additional height and you will greatly increase your radio's capabilities. (Best away from trees but the added height will help greatly).

For using the radio at home, you can greatly enhance your signal just by connecting an external antenna and getting the antenna outside and higher in the air.

So, You Bought a Baofeng Radio – Now what?

Myself I have a 23' painter's extension pole from my last housing renovation project that I use. I will set it up outside with the antenna on it, and a length of coax cable running to my radio. I have instantly increased the range and quality of my signals in and out with a simple solution I can put up or take down as needed.

Unfortunately, I am in a Homeowner's Association community that would have a fit with a big bright yellow painter's pole sticking up in the yard, but that is a discussion for another book.

Note – While this type of antenna works great for a VHF/UHF antenna, placing something unbalanced on it like a directional YAGI antenna will make the pole unbalanced needing guide lines to steady it.

Many folks who live in HOA controlled neighborhoods will place an antenna in the attic. This is usually by folks who are using base radios or even higher wattage mobile radios as a base station, but there is no reason you can't do this with your handheld radio also. In my own area I have no problems reaching the nearby repeaters with a home mobile radio that I use as a base and a mag mount antenna I use near a window in my office. The additional power and close proximity I have to most repeaters makes going to the trouble of an attic installation unnecessary. There is one repeater that I work with on a net weekly that this doesn't work for, but that is a 30-minute net once a week, so I simply work that net from my vehicle outside with my vehicle 50-watt mobile radio. My Baofeng would easily receive from that repeater, but my transmission would be scratchy and since I lead that net, I need to use a stronger radio than a handheld for a cleaner signal.

So, You Bought a Baofeng Radio – Now what?

A Note about SWR

When attaching an external antenna, you should look at using an SWR meter to take a reading of the power out to the antenna. A properly matched antenna will enhance the radio while an antenna not matched properly will actually be a detriment. The SWR reading should be as low as possible. If you join in with a club, the club members are excellent resources to show you how to use one of these meters and can probably loan you one saving you the cost of purchasing one. In addition to the meter itself you will likely need some connector adapters for your radio, so if you borrow or purchase make sure you have the right connections. In Appendix H we will take a look specifically at models of SWR meters, connectors and touch briefly on using an SWR Meter.

My Simple Antenna Solution:

As I have mentioned, I do use simple Painter's extension poles as antenna poles. This was a convenient solution that came in part after the completion of a painting project following selling of my last house. All of a sudden, I had several painters poles and this was the perfect use for them.

For the base, I dug into my Christmas decorations and used a Christmas tree base. In the photo here to the right you can see my outdoor antenna – a 12' 3-part extension pole which will raise my radio horizon to just under 8 miles.

So, You Bought a Baofeng Radio – Now what?

Painter's Pole mounted in a Christmas Tree Base

This painter's pole extends 12' into the air – Radio horizon of 7.88 Miles

To attach the antenna to the pole, I use a paint roller handle cut off and inserted into a 1.5" short length of PVC pipe. To help the paint roller handle fit tightly, I simply wrap it with Duct Tape.

The Antenna mount then attaches to the PVC with two metal rings.

This simple antenna and pole attached then to my Baofeng gives me much greater signal to and from repeaters.

SWR – Standing Wave Ratio

SWR is the method of testing and tuning your radio with the antenna to ensure that you are getting the best performance from the radio. A good

So, You Bought a Baofeng Radio – Now what?

SWR meter will also give you a reading of the power your radio is putting out. For more detailed information on SWR and SWR Meters, look at our Appendix H in the back of this book.

So, You Bought a Baofeng Radio – Now what?

Chapter 8: What Frequency do I use?

Now that you have your license, what frequency should you use?

Well let's get the simple stuff out of the way first.

- For VHF 2M, the SIMPLEX Calling frequency is 146.5200 FM
- For UHF 70cm, The SIMPLEX Calling frequency is 446.0000 FM
- For VHF 1.25, The SIMPLEX Calling frequency is 223.500 FM

So – you run outside, plug in the frequencies, and start calling out to see if your heard. Quiet. Nothing. No response.

Remember the limitations of your radio – Your likely going only 2-3 miles, Maybe up to 5. Your radio horizon when standing with an antenna 6.5 feet up, to another person 6.5 feet is about 7 miles. Add trees, leaves, buildings, hills, moving cars, birds, etc., you're going to get much less. So how do you find a frequency to talk to someone on?

Relax – There's an app for that. Well, there is, and some good websites. But let's review one quick topic before we mention them, so you understand the difference between using your Baofeng in SIMPLEX mode and in using it in DUPLEX or REPEATER mode.

SIMPLEX mode is where you transmit and receive on exactly the same frequency. You transmit on 446.0000 and your buddy hears you. He then responds on 446.0000 and you are listening on 446.0000 hearing his response. This works great for radio-to-radio transmission in close proximity, and it also works great for SIMPLEX repeaters (We will talk about these in chapter 9). But – It doesn't work for most repeaters which are DUPLEX repeaters. For these repeaters there are additional

So, You Bought a Baofeng Radio – Now what?

programming settings you must make in your radio and we will cover those in the sections for CHIRP programming and Hand Programming.

DUPLEX or REPEATER mode is when your radio is configured to actually transmit on one frequency and receive on a slightly offset frequency. We will get into the details about this in chapter 9 when we cover repeaters, but essentially this is done so that a fraction of a moment after the repeater starts to receive your transmission, it can then retransmit you message while you are still talking out on another offset frequency. This allows you to appear to be coming across "Live" on the repeater. A person elsewhere can then key up instantly after your long two-minute message and begin responding. Again, more detail in chapter 9.

Repeaters also use "Tones" that are sub-audible frequencies that YOU transmit on your Baofeng (Or other radio) that triggers the repeater. This lets the repeater know you are INTENTIOANLLY talking to it because you just passed the secret code-handshake – The proper tone. These tones are not actually secret, but by having the right tone then you are likely intending to talk to that repeater so it will respond by passing your message. You can find information on setting these tones in Chapter 3, menu items 11 & 13, R-CTCS and T-CTCS tones.

Why is this important for this chapter?
It is important because as you start planning for frequencies to talk through, both SIMPLEX and DUPLEX Repeater channels, you need to realize that just programming a frequency is not going to be enough. If you just plug in a frequency and start talking, and nobody responds or hears you, then you're going to get confused. So, when talking to a repeater – You must send the proper tone with your message to tell the repeater to let others hear you and repeat your transmission, and you must program your radio for the proper offset frequency.

So now you ask – Where do I find these frequencies?

So, You Bought a Baofeng Radio – Now what?

Where else – On the Internet. There are several sources that you can go to for a list of repeaters. The following list will guide you to those sources.

Online sources for Repeater Listings:

Radio Reference
https://www.radioreference.com/
Radio reference is a site that provides lists of Ham, Commercial and Public frequencies for repeaters and for organizations. It is a tremendous resource where you can find information based on state, county and organization.

RepeaterBook.com
https://repeaterbook.com/
Repeaterbook is focused on lists of HAM repeaters across the country and organizes by state. If your looking for busy repeaters, then when searching for your state look for "Linked Systems" which is where multiple repeaters over a geographic area are connected for wider coverage. These tend the be the busier repeater networks and will offer more chance for contacts. In Florida, the SARNET repeater network covers the entire state allowing one to talk on the net from the keys all the way up to Pensacola – A 16-hour straight drive.

LiveCentral.com
http://www.levinecentral.com/repeaters/google_mapping.php
This is an independent site that provides lists of repeaters available across multiple bands.

So, You Bought a Baofeng Radio – Now what?

Amateur Radio Newsline
https://www.arnewsline.org/repeater-list
This site is a "News Site" for amateur radio and contains an active repeater list.

CHIRP
https://chirp.danplanet.com/projects/chirp/wiki/Home
Chirp's own query capabilities will allow you to query the internet for repeaters based on your county or zip code, then you can add those repeaters to your own radio. You can find out more on this in our chapter on CHIRP later.

Using these resources, you can search for repeaters that are local in your area. I use the first two frequently. You can also use CHIRP if you program your radio with CHIRP, or even just use the query functions within CHIRP to pull a list of 2-Meter and 70-CM repeaters within a specific distance from you. More on that in our CHIRP Chapter.

Now – How do you get these frequencies into the radio?

Two methods – You can program the radios by hand with the proper offset and tones as we will explain in chapter 11, or you can program the radios using CHIRP as we will explain in chapter 10. The benefits of using CHIRP to program your radios include much faster programming and being able to image and back up your radio, and also you can make direct queries from the Repeater Book website and download repeaters based on a "Political" or a "Proximity" query. Basically, Political means you are selecting based on State/County combination, whereas Proximity allows you to select repeaters based on a mileage distance from a zip code that you plug in.

So, You Bought a Baofeng Radio – Now what?

I have found that using CHIRP is by far the easiest method to program these radios. Instructions on using CHIRP are provided coming up in chapter 11.

So, You Bought a Baofeng Radio – Now what?

… So, You Bought a Baofeng Radio – Now what?

Chapter 9: Talk About Repeaters

We have already talked about repeaters several times. But now we want to have some focus on these wonderful devices and make sure you understand just how important they are.

I live in Florida. Between June and December, we live in Hurricane country – always aware and ready just in case. (August through October are the most active months). For Ham operators, there are two types of Ham use around hurricane season. HF Radio allows an operator to set up anywhere and using either a traditional HF radio antenna, or what is referred to as an NVIS antenna for short-range HF, a ham operator can communicate without the need for repeaters. (NVIS is Near Vertical Incidence Skywave)

But – add a repeater network and now we immediately increase our local-use capabilities and we grow our pool of volunteers. Most of the shelters around the Tampa Bay area where I live are within range of a repeater with a handheld radio. And during Hurricane Michael in 2018 when the panhandle was devastated, all of the repeaters stayed up and operational allowing hams to still communicate from their handheld radios even though cellular systems were devastated for weeks. Line cuts had effectively ruined most cellular networks for over two weeks except for one.

Full Duplex Repeaters are electronic frequency repeating devices that receive a signal on one frequency and retransmit on a second frequency. Usually these frequencies are slightly offset. In the VHF band the normal offset is 600 kHz + or – from the primary frequency, and in UHF the offset is 5Mhz + or - from the primary frequency. This means that if you program your radio to receive on 449.125 MHz with an offset of (-) 5Mhz, then the radio will transmit on 444.125. This offset allows the

So, You Bought a Baofeng Radio – Now what?

repeater to listen on one frequency, 444.125Mhz and then immediately retransmit on 449.125 to all other operators. You won't hear the transmission of course because you have your microphone keyed and are talking, but all other operators will hear you. This is what allows for "Live" communication – no delay between your speaking and others hearing you.

> *** TIP:*
>
> *If you have your Baofeng programmed for a repeater on a channel, and your display is set to display the frequency rather than the name of the channel or the channel number, then you can use the [*SCAN] button to SWITCH the repeater TX and RX frequencies. This can be used to troubleshoot a radio configuration – you will actually hear others talking to the repeater THAT ARE WITHIN YOUR LISTENING RANGE, but you will not hear the repeater. More about this later in troubleshooting.*

In addition to having offset frequencies, you also will have to program your radio with a tone to use as a "Squelch Tone". This is a tone that the repeater specifically listens for to know to repeat the incoming signal. There are a number of standard CTCSS Squelch tones that your radio can be set for and which tones to use are depending on the settings of the repeater you are using. (Radio Menu items 11 & 13 cover these tones)

CTCSS RADIO TONES

67.0	82.5	100.0	123.0	151.4	186.2	225.7
69.3	85.4	103.5	127.3	156.7	192.8	229.1
79.9	88.5	107.2	131.8	162.2	203.5	233.6
74.4	91.5	110.9	136.5	167.9	206.5	241.8
77.0	94.8	114.8	141.3	173.8	210.7	250.3
79.7	97.4	118.8	146.2	179.9	218.1	254.1

So, You Bought a Baofeng Radio – Now what?

In most cases these tones will be published by the club or organization that runs the repeater for public use. But – If the club or organization chooses to not publish the tone that is their option. Some more advance radios (More so that then Baofeng) can automatically detect and set themselves for the specific PL tone being used. But for your Baofeng – You will have to depend on the published setting.

If you are programming from CHIRP and perform a Repeater Book query using chirp, in nearly all cases the proper tone will be pulled and programmed for you.

Repeater Planning

The frequencies used by repeaters in an area will be coordinated by a Repeater Coordinator, which is an organization that tracks and plans all repeaters in a geographic area. It is the responsibility of the Repeater Coordinator to track the power, location and frequency of all repeaters as well as the tones being used, and then to assign out frequencies and tones as well as any other information to repeater operators wishing to stand up a new repeater. This coordination is meant to prevent conflict between repeaters. Frequencies and settings are reused in an area normally. In Florida we have many repeaters on the same frequencies and tones, however the repeater coordinator ensures that these are not located within a close proximity of each other thus preventing confusion and communications overlaps.

In the Appendix you will find a list of most of the Repeater Coordinators in the US. Some are state based, others are regional.

The electronics of the repeater are located within a data center or communications center cabinet. The repeater is connected to antennas which are mounted as high as possible. Popular locations include the tops of tall buildings, or on antenna towers running several hundred to

So, You Bought a Baofeng Radio – Now what?

over a thousand feet in the air. These are shared towers where multiple antenna services will run from including community Television and Radio. The higher the better for us because the repeaters we leverage are VHF/UHF FM repeaters – Line-of-Sight. The greater the height, the greater the range we can talk.

In chapter seven we mentioned four online directories for finding repeaters. The best way honestly though is through using CHIRP covered in chapter ten coming up and performing a Proximity or Political query of Repeater Book's website through the CHIRP interface. CHIRP will tell you what repeaters are nearby, but they really won't tell you much more. To get more details about specific repeaters or repeater networks try looking for them on one of these websites:

1. Radio Reference
   https://www.radioreference.com/

2. RepeaterBook.com
   https://repeaterbook.com/

3. LiveCentral.com
   http://www.levinecentral.com/repeaters/google_mapping.php

4. Amateur Radio Newsline
   https://www.arnewsline.org/repeater-list

5. CHIRP
   https://chirp.danplanet.com/projects/chirp/wiki/Home
   I am including Chirp because you can run the Query function in Chirp and pull a list of repeaters based on county or distance from your location. That list can be exported for use later and even printed.

So, You Bought a Baofeng Radio – Now what?

Repeater Cost

If you are wondering about how much it costs to get a **Professional** Full-Duplex repeater up and running, the cost usually starts at a couple of thousand dollars and heads up from there. Repeaters are sophisticated hardware that go well beyond that of your Baofeng. The filtering for a full-duplex repeater is performed by a high-quality Duplexer which makes it possible for the repeater to be able to filter the received signals and transmitted signals for simultaneous work on the same antenna system.

If you are wanting to play, there are some adapters you can purchase to allow you to connect two radios together and configure them as repeaters. These are definitely "Amateur" repeaters and may work for temporary use but would not make a good permanent solution. See the end of this chapter for more on this.

SIMPLEX Repeaters

If you do want to play with repeaters at an entry-level stage, there is an inexpensive repeater solution that you can get into for around a hundred dollars that may interest you. In this case, enter the SIMPLEX Repeater. These are quite simple devices in comparison to the complex Full Duplex repeaters that we have been talking about, and function much more simply.

With a simplex repeater, a small device is set up with a radio. The radio in fact can even be your Baofeng. The device is attached to the radio through a cable similar to our programming cable, but instead of it being a USB cable, this is a form of audio cable that runs from the Kenwood Mic/Speaker connector on your radio into the small simplex repeater, often just a bit bigger than the radio itself. The radio is placed on a frequency in Simplex mode – same receive and transmit frequency, and then the repeater is plugged in. When the radio receives a communication from someone, it simply records the message coming in and then when finished it re-transmits the message out on the same radio.

For you – the person sending the message, this is odd because immediately after you speak your message into the radio, then the repeater plays back your message in your voice. For others that are within hearing distance of your radio they will also hear your message twice – the first time when you originally sent it and the second time when the message goes back out. But for anyone else not in range of your radio they will only hear you once.

Just like a Full-duplex repeater, you can set your radio and simplex repeater up with a raised antenna which improves the transmission range. I have two types of antennas I use for this purpose – One is a military style pole tower with eight 4-foot fiberglass poles that can be raised up to 32' in the air. I also have then a lightweight 8' painters' pole that I can attach for a total height of 40', then the antenna is added. Based on the formula we covered in chapter three about radio horizons, this extends my radio range out as follows:

$$\text{Radio Horizon (In Km)} = 4.12 \times \sqrt{\text{Height (In Meters)}};$$
$$\text{and } 40 \text{ Feet} = 12.2 \text{ Meters}$$

$$\text{Radio Horizon (Km)} = 4.12 \times \sqrt{12.2}; \quad \text{and } \sqrt{12.2} = 3.49$$

$$\text{Radio Horizon (Km)} = 4.12 \times 3.49; \quad \text{therefore}$$

$$\text{Radio Horizon (Km)} = 4.12 \times 3.49 \text{ which equals } 14.38 \text{ km}$$

$$14.38 \text{ km} \times .621 = 8.92 \text{ Miles}$$

So, You Bought a Baofeng Radio – Now what?

So, if looking for a way to deploy a short-term emergency repeater, I can actually do this relatively easily with my Baofeng Handheld radio! Now this will require some cable and antenna adapters to get the antenna up and operating from our radio which we will cover in a later chapter.

Besides an external device to have for working as a repeater, I need to also mention that you can use some models of radios as simplex repeaters without the need for any additional hardware. Though this is a capability that is beyond the capabilities of your Baofeng radio, I do happen to own a more advanced radio by Baofeng Tech which is their DMR 6x2 radio. Though this is a DMR Digital radio, the radio also works fine as an analog radio. Advantages of this radio include being able to store 4,000 channels, being able to group channels together, and having advanced functions such as a simplex repeater built in. That's right – I can set this radio up without anything additional as a small simplex repeater.

If you decide you would like to play with setting up a simplex repeater, two models which I am aware of are:

- SURECOM makes several types of small repeaters. All will require one to two radios for use for receiving and transmitting. The models they make are in both DUPLEX and SIMPLEX.

- ARGENT makes a small simplex repeater that can be used with a single radio for simplex repeater operation. I have had better success with this repeater for SIMPLEX operation.

You will find that for a Technician level Ham operator, that repeaters are the best tool you can to expand your range and use of your new Baofeng radio. If you do fall into the Prepper community, then a simple simplex repeater may be something that interests you for extending your radio's operational range.

So, You Bought a Baofeng Radio – Now what?

Building a simple Duplex Radio repeater with two Radios

If you take two Baofeng (Or other manufacturers radios) you can connect them together to form a small repeater for playing with. If you select to do this, then do it with the radio wattage set to a minimal (1 watt) and make sure you are working on two OPEN SIMPLEX frequencies – do not do this with just any frequency.

The quality of the repeater function will depend on several criteria, including antennas used, antenna separation, and radio separation. Also remember that these are actually relatively inexpensive and lower quality radios, so performance quality will vary.

The parts that you generally will need to assemble a simple duplex dual-Baofeng Radio repeater are:

1. (2) Baofeng Radios
2. (2) Antennas – Quality of the antennas will make a huge difference in how well this will work. Suggest at least (2) Mag-mount or external antennas.
3. A SainSonic RPT-2D two-way repeater box (Amazon for about $30)
4. (2) Additional radios for TESTING your repeater

Here are the quick steps to set up a 2-radio Baofeng Repeater, along with a link to a good video on YouTube. Both radios will be set up ALMOST the same way with the exception that one of the repeater radios will have the VOX menu setting turned on with the other turned off.

So, You Bought a Baofeng Radio – Now what?

1. Identify one radio as "REPTR-TX" Radio and the other as "REPTR-RX".

2. Use the following settings for BOTH radios:

Menu Option	Menu #	Setting
TXP (Transmit Power)	2	Low
R-CTCS (Receive CTCSS)	11	250.3 Hz
T-CTCS (Transmit CTCSS)	13	250.3 Hz
SFT-D (Shift Offset Direction)	25	Negative (-)
Offset (Value of the Offset	26	000.600

Notes:
(1) We are setting Power to "Low" for testing purposes. On the Transmitting radio this can be set to High for better range.
(2) The R-CTCS & T-CTCS values can be any of the values available. This setting is just an example.

3. For your REPTR-TX Radio set the radio in FREQUENCY mode and set the frequency to **146.900**. This will transmit out the messages being forwarded. Make sure you are not using an offset value or a Shift Offset Direction value.

4. For your REPTR-RX Radio set the radio in FREQUENCY mode and set the frequency to **146.300**. This will receive in the messages being forwarded. Make sure you are not using an offset value or a Shift Offset Direction value.

5. <u>**Set the power on both radios to 1-watt to limit range for testing.**</u> (We are testing and don't want to create interference).

6. For both radios set Menu option 11 - the R-CTCS (Receive CTCSS) tone squelch to **250.3 Hz** (You can set to any value but use the same value for ALL radios).

So, You Bought a Baofeng Radio – Now what?

7. For both radios set Menu option 13 - the T-CTCS (Transmit CTCSS) tone squelch to **250.3 Hz** (You can set to any value but use the same value for ALL radios).

8. On the TRANSMITTING Radio, REPTR-TX Radio, you will need to now adjust the VOX Setting to **10**. This allows the radio to automatically transmit anything coming in through the Microphone (At a high enough volume level).

9. You will need to connect both radios together either with the SainSonic Repeater Box – This box basically connects the speaker/Ear jack of the receiving radio to the microphone jack of the transmitting radio.

   Note: I have seen some examples of just using a 2.5mm to 3.5mm connector cable – but I have not had success with this. The SainSonic is an inexpensive $30 device that has worked well for me.

Now that we have our repeater radios set up and connected, we now need to set up our test radios to work with the repeater. The repeater has been set up to use a CTCSS for Receive and Transmit of 250.3 Hz. This means that with our test radios, we will HAVE to transmit that specific tone to trip the squelch of the receiving radio and for it to then, through our interconnected cable transfer our audio to the second radio which will then broadcast it out. The PTT button will NOT have to be pushed because we have set our VOX setting to transmit with it hears a communication on the microphone jack connection.

So, You Bought a Baofeng Radio – Now what?

Both of your test radios will have to have the following settings for them:

Menu Option	Menu #	Setting
TXP (Transmit Power)	2	Low
R-CTCS (Receive CTCSS)	11	250.3 Hz
T-CTCS (Transmit CTCSS)	13	250.3 Hz
SFT-D (Shift Offset Direction)	25	Negative (-)
Offset (Value of the Offset	26	000.600
Frequency set to for receive*	Freq Mode	146.900

* With SFT-D and Offset set, the Transmit frequency will automatically be set to 146.300

For BOTH radios set them as follows:

1. Set the radio to FREQUENCY MODE and set the radio frequency to **146.900**.

2. Set the R-CTCS (Receive CTCSS), Menu #11, to **250.3** Hz. If not set your radio will NOT hear the transmitting radio in the repeater.

3. Set the T-CTCS (Transmit CTCSS), Menu #13, to **250.3** Hz. If not set you radio will not TRIP the receiving radio and it will not pass the communication over to the transmitting radio.

10. On both radios we now need to go to FREQUENCY SHIFT which is Menu 26 and set our offset value to **00.600** (**600 kHz** for VHF. For UHF this value would be 05.000, or 5MHz).

11. On both radios we now go to SFT-D which is our Shift Direction, Menu #25 and set direction to (-) for Negative.

So, You Bought a Baofeng Radio – Now what?

What will happen in our little repeater setup is the following:

- Operator "A" sends a communication through his radio. This radio is set on frequency **146.900** for RECEIVE, but with the offset set to a **-600kHz** it actually transmits on **146.300**.
- Repeater Receiving radio picks up the communication and plays it through the "HEADPHONE" jack which is connected to the "MICROPHONE" jack of the second radio. The repeater receiving radio is set to **146.300 SIMPLEX** with no offset or shift.
- The TRANSMIT radio of the repeater pair is set to VOX allowing any input, which is coming through the Microphone jack, to be transmitted WITHOUT having to key the PTT key.
- The TRANSMIT radio now transmits the signal coming in from the Microphone jack. This radio is transmitting on **146.900 SIMPLEX** (No Offset or Shift).
- Operator "B"'s radio, set up the same was as Operator "A"'s radio is set up – Receiving on frequency 146.900 and picks up the transmission playing it over the radio's speaker.
- With both Operator radios set the same, Operator "B" can respond back and the repeater will retransmit simultaneously so Operator "A" can hear it.

So, You Bought a Baofeng Radio – Now what?

- If the two antennas for the repeater are set up high enough then your range is extended so that both operators can be much farther apart and still hear each other.

A few points to remember about this repeater – It has limitations.

- Make sure you check Frequencies with the Frequency council in your state or area, so you are not transmitting on any frequencies already in use.
- The R-CTCS & T-CTCS settings, Baofeng menu options 11 & 13, are to prevent the repeater from transmitting unwanted messages. This would need to be set up on the Repeater radio pair and, on any radios, using this repeater. He value of 250.3 Hz we are using is just one of many values you can use.
- Consider upgrading the batteries to long lasting batteries with a USB to Battery connector on the battery, then use a 12-volt battery with a USB Port adapter on it to give the repeater longer use life.
- If using this repeater, I would try to house it in a water-proof container – Such as a plastic Ammo box that will stay dry in the rain. Also consider covering it as well.
- Consider use of externally connected antennas for better reception & transmission.
- Consider using either an Amplifier on the transmitting radio such as one of the Baofeng Tech AMP-25 amplifiers which will boost your signal to a max of 40 watts out.
- Remember Baofeng radios have a notorious reputation for harmonics – If setting this up you might consider a higher quality radio without that problem for transmitting that will not transmit harmonic frequencies.
- The Baofeng Radios are low-quality radios with short duty cycles. This means they are not designed for frequent and long transmission

So, You Bought a Baofeng Radio – Now what?

times. At a high power you could burn up the transmitting radio faster than otherwise normal for these radios.
- Our little transmitter here requires TWO antennas – We are not using a Duplexer which can split signals coming in and out. Both antennas should be spaced apart if possible.
- It is possible to use this repeater as a Cross-Band repeater – Receive on VHF/UHF and transmit on the other band. This is a simple setup for the repeater radios. For the Handheld radios – Your setup would vary. Set channel "A" for transmitting on the frequency the repeater is listening on and set channel "B" for receiving on the frequency the repeater is broadcasting on. Then make sure your TDR setting (Menu Option 7) is set to ON so that the radio monitors on both frequencies. You will NOT want to set your SHIFT or OFFSET values as you will now be transmitting and receiving on SIMPLEX.
- The next step would be to add a VHF/UHF Duplexer which allows a single antenna to be used with two radios - if one radio is on the VHF band and the other the UHF band.

I have found this Baofeng Duplex Repeater project a very interesting project for learning about repeaters. I have built one myself for use in a Ham Radio and Baofeng class which combined with a pair of 23 foot Painter's poles will give me a simple solution allowing me to talk in an area about 7 miles from my established location, that is 7 miles in either direction for a maximum range of 14 miles. Factor in interference from trees and nature, I should still be able to get 4-5 miles easy from my base, or 8-10 miles diameter.

So, You Bought a Baofeng Radio – Now what?

> **TIP: Repeaters and the [*] keypad key**
> One feature of the Baofeng Radios that I want to also add here – When talking to a repeater in Repeater Mode (Off set established), you can press the [*] key on the keypad to flip the transmit and receive frequencies. This can be used to "Hear" what a repeater is actually hearing for troubleshooting purposes.

Definitions: SIMPLEX, DUPLEX, and Cross-Band Repeaters

SIMPLEX Repeater: This is a small device that connects to your radio and records incoming messages digitally. The message then is immediately re-broadcast back out on the same radio that received the signal.

Characteristics include:
- Cheaper (1 Radio, 1 Antenna, 1 Repeater box)
- Simpler – Works well with SIMPLEX, no Offsets or Shift settings
- Users have to listen to their own message after transmitting, so better for short messaging back and forth.
- Works with all HAM radios and can work with GMRS Radios as well.

DUPLEX Repeater: This type of repeater has two components – Receiver and Transmitter. In a small device such as we built it requires two radios and can require two antennas. A Professional quality Duplex repeater can cost $1,000 or more – These are the most common types you will find readily available to connect to.

Characteristics include:
- More expensive and complex than a Simplex repeater.
- More equipment.

So, You Bought a Baofeng Radio – Now what?

- Can work with 1 or 2 antennas – With 1 Antenna you will need a Duplexer.
- Filtering requirements for single-band 1-antenna repeater are for high quality – high-quality duplexer will be needed ($$$).
- Works with all HAM radios and can work with GMRS Radios as well as long as the radios support CTCSS, Shift Offset, etc)

Cross-Band Repeater: This repeater is a DUPLEX repeater however it uses two separate bands for Receive and Transmit. By using two bands we are receiving and transmitting with significant frequency separation allowing Duplex repeater functions to occur in simpler radio systems.

Characteristics include:
- DUPLEX capabilities with less expensive equipment than a professional DUPLEX Repeater.
- Many Mobile transceivers can perform this function.
- Easier to work with 1 antenna.
- Radios have to support Dual-Monitoring (As the Baofengs do) to operate on two frequencies at the same time. Single-receiver/channel transceivers will not work with a cross-band repeater.

What happens when the power or Internet goes out to Repeaters?

In the event of a wide-spread power outage what happens to a repeater or a repeater network? Most amateur ham repeaters are run by clubs. Most are also set up with some type of backup power – either a UPS for a short-duration power outage, or with a Generator to provide longer duration power since a generator can be refueled.

But the establishment of backup power is totally up to the repeater operators. In a repeater network, you will likely have a good backup power plan established. As a part of the network, the repeaters are more

So, You Bought a Baofeng Radio – Now what?

critical and noticed in the event of a failure. A responsible repeater owner participating in a network will likely take this into account and provide a backup contingency plan.

But what about if communications or the Internet goes down? For independent repeaters, loss of communications or the Internet will not make too big of a difference. Independent repeaters work alone where they extend the range of VHF/UHF radio communications. But – If the repeater is a part of a multi-repeater coordinated network, and if the repeaters are connected via a back-bone Internet connection, then in most cases the repeaters will drop into an "Independent" mode where they still work, but each as an independent repeater covering a very localized area.

In my area of Florida, there is a very well-known multi-repeater network heavily used for casual communications as well as emergency planning and emergency coordination. This repeater network has an Internet backbone for synchronizing all repeater transmissions across the area. If a failure in the Internet does occur, then repeaters not able to connect will then work as independent local repeaters.

The reason this is important to understand is so that you know that the dependence on a repeater for extending your range works only as well as the current situation. Repeaters are wonderful devices but factor the possibility of not having access to these into your emergency plans. Without access to a repeater, your next best solutions will be:

- Assembling and readying an external antenna, best on a pole, for extending your radio range.
- Purchase and set up a 25-watt or 50-watt mobile radio with an appropriate antenna for more power and range.

So, You Bought a Baofeng Radio – Now what?

- Consider taking your license to the next level – A General license allows you to get a HF Radio which allows for communication over much greater distances (Hundreds to Thousands of miles).

Chapter 10: Radio Nets and Finding them

A Radio net on Ham Radio is a talk group on the radio. These nets are either organized CONTROL nets where there is a Net Control Operator who manages the communication on the net, or they are OPEN informal nets where any participant can jump in and communicate. Nets are performed for a variety of reasons ranging from simple conversation, to passing of information, or during an emergency event they are used for critical communication between areas that may have no other working method of communication, and the passing of emergency and tactical information.

In Florida and many areas that are prone to natural weather events such as hurricanes or flooding ARES and RACES act as two organizations that regularly operate nets for practice or real-world activity. I am myself a NET Coordinator for our county ARES organization and we perform weekly nets to allow participants to check their radios and to make announcements for activities and events that may be occurring throughout our area.

As a licensed Ham operator with a handheld radio, you can join a net during its operation. Nets are very often used to allow radio operators to "Check In" to the net and provide announcements and news to the operators listening. If you don't want to check in that is fine – but checking in does allow you to see how your radio is functioning through the particular repeater you are talking on and even to try various antenna configurations or location configurations. And check in on multiple nets through multiple repeaters to get a feel for how well your radio works from various locations. Doing so allowed me to understand that my radio does perform generally well from areas 30 to 40 miles from repeaters, and that for some repeaters I need to be located closer

So, You Bought a Baofeng Radio — Now what?

for a good signal repeat. This also gets you familiar with other radio operators around you and allows them to get familiar with you.

CONTROL vs OPEN Nets

There are two types of control for the nets — Control and Open. In a control net the NET Control Station (NCS) or Net Control Operator (NCO) is the person who acts as a traffic cop on the net — acknowledging individuals checking into the net and passing communication. On a large net this coordination can be critical to keep traffic flowing smoothly on the net. An OPEN Net is a casual and much less formal format with no specific NCO running the net. This works best for net with a smaller group of participants and when not trying to pass critical information. Some of the types of situations where a CONTROL net may be used include:

- **Traffic Nets** — Used for passing of written messages
- **Resource Nets** — Used during an event by incoming operators and volunteers to check in and receive assignments, or to locate needed equipment, supplies, operators or other types of resources.
- **Tactical Nets** — Used for emergency traffic and coordination
- **Information Nets** — Used for passing announcements and dissemination of information allowing two-way communication.
- **Health & Welfare Nets** — Handle messages between friends and family usually in a disaster area to pass status information.
- **Command Nets** — Used in all large disasters or emergencies for top leadership for passing top-level critical information.
- **Specialized Nets** — Set up usually for agencies or organizations such as the Red Cross, Salvation Army, the Weather Service, or other similar organizations for passing unique information pertaining to them.

So, You Bought a Baofeng Radio – Now what?

Finding a Net

So how do you find a net to participate on? Through your local clubs and through their websites you will usually be able to find a list of Nets and what repeaters they run on. You will want to locate a club in your area because they will be on nets and repeaters that are within your area. The best initial source for clubs to find will be through the ARRL's website and in their Affiliated Clubs page at this link:

<u>http://www.arrl.org/find-a-club</u>

Note that you may have to be a bit broad in your search criteria to find a nearby club. As you find clubs near to you, go to their website and look for a section related to "Weekly Nets". In the club's Net listings there should be a schedule of the club nets along with repeater details with Frequency, PL-Tone, time and days.

If you make a note of the repeater frequencies and club name, then you may be able to easily locate the repeater using CHIRP and program it into your radio as explained in the next chapter, *Programming with CHIRP*. Searching for a net based on your club or an organization will be the best first-choice so that you can find a net based on a specific interest.

Another easy source to look for a NET is also from the ARRL in their NETS Directory – An online listing of registered nets.

<u>http://www.arrl.org/arrl-net-directory-search</u>

Here you can find nets based on geographic location, radio band, and general affiliation. This will be a good source to find emergency nets that may be in your area.

So, You Bought a Baofeng Radio – Now what?

If you are looking for an emergency planning type of net, you can also check out the ARRL ARES groups near you by visiting the following website page and looking up the section nearest you. By visiting the local chapter website, you should be able to find information on NETs, events, and even meetings that are local to you.

http://www.arrl.org/sections

Through the ARRL West Central Florida page for my area, I can find the section's own website, Facebook and Twitter pages, and details about what is happening in my area. Here in this area of Florida we use a Networked repeater system heavily which hosts sectional nets, Sky warn nets, and several other repeater nets that cover a ten-country area in West Central Florida. This is an area of over 9,000 square miles which you can talk on!

Talking on a Net

Many times new hams will suffer from "Mic Fright" nervous to talk on a net. At the most basic, checking into a net is a very simple passing of basic information including your call sign, first name (Not last), general location and club affiliation if you belong to a club or organization. Nets are often run from scripts which allow multiple NCOs who alter running the nets to follow the same or very similar script each time the net is run. This produces familiarity on the net which is a very good thing – It is best to keep each net running in a very structured format.

So how does the conversation on a net flow? On the following page we are going to provide a sample script to give you a quick idea. The script will be from two people on the net, the first being NCO and the second being HAM.

So, You Bought a Baofeng Radio – Now what?

Sample Net conversation requesting check-in and simple check-in response between Net Control Operator (NCO) and Ham:

> *NCO: "This is Net Control. We will now proceed with General Check-Ins. WE ENCOURAGE AND WELCOME ALL operators to check in."*
>
> *NCO: "Any stations wishing to check into the Hillsborough ARES RACES Net, please call now one at a time."*
>
> *NCO: "Please give your call sign, name, and location. General Check-Ins, please call now."*
>
>   HAM: "Net Control, This is CALL-SIGN, Joe in Tampa, Checking In."
>
>   "No Traffic or Comments for the net."
>
> *NCO: "Joe, Thank-you for checking in. I have you in the log and you have a good signal. "*
>
> *NCO: "Next Station, please call now"*

That's really all there is to checking into a net. Your giving some basic information about yourself and announcing if you have anything to say on the net. Depending on how the net works, the NCO may take your comments or questions during check-in or may ask you to hold your comments until after all check-ins, and then may come back to you afterwards.

So, You Bought a Baofeng Radio – Now what?

SIMPLEX Nets

Some groups will on occasion run SIMPLEX nets to test simplex operation for an area. This can be an important net because it truly does test the range capabilities of the radios that operators have and your ability to operate without repeaters. Remember that on a hand-held radio, you will likely only have a two to three-mile range at most. If you are in a mountainous area your range will go farther, extending to several miles but still only line-of-sight (Not through mountains).

With a handheld radio your best communications option to extend your range will be to use an external antenna – maybe a car-mount antenna placed up high to gain additional distance. Raising an antenna 15 feet into the air will get you just about 5.5 miles of range and getting it up to 30 feet around almost 8 miles of range (Perfect conditions).

In a Simplex Net you will have multiple operators working their radios at the same time and you will be able to hear only some of them. The idea is to be able to be heard by someone who in turn is heard by others, and therefore being able to relay to someone else. In a disaster situation your being able to have a message relayed or relaying for others could be the only way of getting your communication out. These types of nets would function as OPEN nets since there would be the lack of a central NCO.

The biggest takeaway from this chapter should be to find some nets – and just check in. Use the nets to confirm your ability to be heard and understand what repeaters you can be heard on. If you have problems with reaching others, simple changes like connecting a car mag-mount antenna to your Baofeng or simply walking outside or upstairs can help you reach the repeater and the net.

In the end – Find the nets around you, check in, and test your gear.

Chapter 11: Programming with CHIRP

Before we even get into hand programming the Baofeng, which you do need to know, we simpler and more common method on programming of Baofeng Radios with CHIRP. The use of CHIRP allows easing programming of the radios giving you the ability to pull repeater lists dynamically, storing radio images, or programming with pre-defined radio images.

CHIRP is available for Microsoft Windows, Apple MAC OS, and for Linux systems. For the purposes of our book, we are ONLY going to be working with CHIRP on a Microsoft Windows computer. Though it does work in the other operating systems I do not myself have experience with it in those environments and cannot assist.

Getting CHIRP
Start by downloading CHIRP from the following website: https://chirp.danplanet.com/projects/chirp/wiki/Download. You will need to download the proper installer and install the software.

Getting the Cable – Baofeng PC-03 Cable

To use Chirp with the Baofeng radios, you need the proper cable. This is because of a chip in the cable that connects with the radio. The proper cable is sold by Baofeng Tech and is their PC-03 cable available by 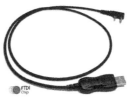 going onto Amazon's website or the Baofeng Tech website at https://baofengtech.com/Programming-cable. This chip contains a FTDI chip and has two LEDs on the cable end that plugs into the computer. These LEDs flash furiously when data is being passed between the computer and the radio.

So, You Bought a Baofeng Radio – Now what?

Why the PC-03 Cable?

This cable works seamlessly with the USB port of your computer and the drivers with Windows to interface the computer to the radio. Many other "Clone" cables will not and will require the installation of a "Prolific 3.2.0.0" driver on your windows workstation to communicate. The problem is that whenever your computer goes through an update then this driver will be overwritten and have to be re-installed. I have clone cables that I used on radios years ago that will no longer work without this less-than-elegant fix and when I purchased my most recent radio a year ago and I picked up the correct cable. The correct cable is a simple and elegant solution.

If you have a clone cable that does not have the FTDI chip, that cable may work with the Baofeng programming software., My experience with that software is minimal due to the fact that I use and strongly recommend CHIRP to program radios. CHIRP works with multiple manufacturers as a more simplified solution. With Chirp you can take a configuration used on one radio and easily copy it to other radios. Myself I use Chirp with by Baofeng radios, Yaesu and TYT Radios so that all of the radios have the same channel assignments.

You will need a good Programming Image

During the programming session we will open an image that I have already set up and made available. From this image you can copy the entire image of channels to your radio or select a group of channels to copy over. At a minimum I would recommend copying the SARNET and NI4CE Repeater net channels over. In addition, you may wish to copy over the Weather or the FRS/GMRS Frequencies. The image I have programmed works for repeaters in the Tampa Bay area.

https://www.SoYouBoughtABaofeng.com/Images/

So, You Bought a Baofeng Radio – Now what?

You can pre-download this image and have it ready to use when we program the radios.

BASIC STEPS

Following are the basics that we will walk through to download your radio image (Even if unprogrammed, this needs to execute).

1. Download and install CHIRP

2. With the radio turned OFF you can connect the Radio to the Computer
   using your PC-03 Cable.

3. Open CHIRP on your computer.

4. In CHIRP select RADIO → DOWNLOAD FROM RADIO.
   This will bring up a pop-up allowing you to select a COM Port. Note that COM PORTS used to be used on older computers. If your computer has a DB-9 COM Port, it is old probably. But – The drivers in the cable and windows allow a USB port to be used though CHIRP will see it as a COM port. Select one of the COM ports, we may have to select different ones on multiple attempts. (In most cases you will only have 1 selection)

5. Select the VENDOR for your radio – Probably Baofeng depending on your Radio, maybe BTECH. Different manufacturers have different radio drivers.

6. Select the Model of Radio.

7. Click OK and follow any pop-up messages. This may include a driver warning. You should get a specific set of instructions for your radio. Follow them and click OK.

8. This step will now start downloading the image of your radio. Even if your radio is unprogrammed, then this blank image will be the

So, You Bought a Baofeng Radio – Now what?

template we will use for uploading back to the radio later. As this step runs, you should see the LEDs on the PC-03 cable flashing and a status bar on the screen while the image is downloaded.

DOWNLOADING REPEATERS FROM REPEATER BOOK

One of the biggest strengths with CHRIP is the ability to download from Repeaterbook already identified repeaters. Note that the repeaters in the list are repeaters that were known to have been running at some point but are not guaranteed to be active now.

After you download your radio image, you will have a template that we will be able to later copy channels to. But to get those other channels, we will need to run a query to get a list of frequencies that can overlay on those channels. Let's do this for 70cm now:

1. In CHIRP select RADIO → QUERY DATA SOURCE → REPEATER BOOK → REPEATERBOOK POLITICAL QUERY
   Note: The difference between PROXIMATY and POLITICAL:

   Proximity allows you to pull a list of repeaters based on BAND a specified distance from a zip code.

   Political allows you to pull a list of repeaters based on BAND a specified on a specific COUNTY.

2. With the Political query – select your county and select the band you want to download for.

3. Click through to continue and you will have a list of repeaters.

4. Just like copying a row in Excel, you can copy specific rows with repeaters and then go to your radio template and paste them into a blank channel, or you can paste over an existing channel to overwrite that channel.

So, You Bought a Baofeng Radio – Now what?

RE-WRITING TO THE RADIO
When ready to re-write to our radio, the steps are actually relatively simple.

1. Select RADIO → UPLOAD TO RADIO

2. Select your radio Brand & Model just as you did in the download steps.

3. Upload your image.

Other Radio Frequency Configurations
We covered the Repeater Book queries which pulled in local repeaters for us to add to our radio. That was easy. What if you want some other channels added? In addition to Repeaters you can also add the following:

- NOAA Weather Alerts (10 Frequencies)
- US Calling Frequencies (4 Frequencies, 1 for 6m, 2m, 1.25m & 70cm)
- US FRS and GMRS Channels (22 Frequencies)
- US Marine VHF Channels (60 Frequencies)
- US MURS Channels (5 Frequencies)

All of these configurations can be pulled from using the option of FILE → OPEN STOCK CONFIG → Select a list of channels.

For instance – To add the NOAA Weather channels, select:
FILE → OPEN STOCK CONFIG → NOAA Weather Alerts

This will open a list of 10 frequencies that can be copied and pasted to your radio image the exact same way we in our repeater lookup and programming steps above.

So, You Bought a Baofeng Radio – Now what?

Remember – You CANNOT Legally transmit on ANY of these frequencies.
BACKING UP YOUR RADIO
Why would you want to back up your radio?

Say you have it programmed with all of the common repeaters you need for Tampa Bay. But – Now you're going camping up in the Mountains. Or just out of the area. And – Your cell phone doesn't work. With Chirp – you can quickly add new repeaters in your area to your radio and you have a way of communicating in case of an emergency – At least if someone is listening. But if you're near a repeater on a network like SARNET then you will be able to get out anywhere in the state.

But – Remember – If you are remote camping – take an antenna or some extra antenna feed line so you could put an antenna 10' or so up in the air for a better signal. If choosing between Power and better Antenna – Go for Antenna.

With your radio backed up when you return home you can just download your radio, open your backup file, copy the channels over with Copy & Paste, and your back! Whole thing will take about 5 minutes or less.

Setting Radio settings with CHIRP for the BAOFENG Radio
In addition to being able to set and program Frequencies with Chirp, you can also manage most of the settings of the radio as well. To do this, you need to be on your Radio tab (After reading from the radio) and look in the upper left area of the screen where you will see a side-tab titled "Settings".

So, You Bought a Baofeng Radio – Now what?

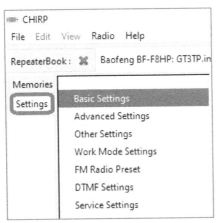

*The **Settings** area in CHIRP allows for changing your radio settings*

Here you can control many areas of the radio including Battery settings, LED Colors, Squelch settings, VOX, Dual Watch, Scan settings, Power-on message, frequency ranges, DTMF settings and much more. You will be able to find screen shots of all of these menus in the Appendix at the back of this book.

In the top image shown on the following pages, we show the options from Chirp for setting basic settings for the radio itself. Some of these settings can also be set through the radio MENU options we discussed in section 3. The lower image for "Other Settings"

So, You Bought a Baofeng Radio – Now what?

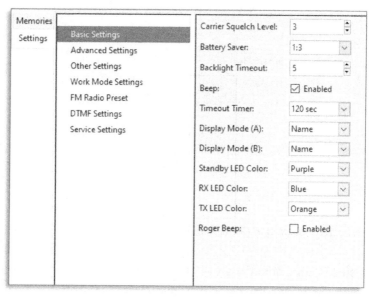

Settings-Basic Settings from Chirp – These screens will be different among different manufacturers of Radio. This screen is for the Baofeng UV-82 series of handheld radio.

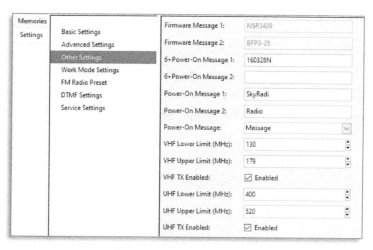

Settings-Other Settings – Here are some settings you can configure that are NOT available through the radio's MENU options through hand programming. Notice Power-On message 1&2, and VHF/UHF Lower & Upper Limits.

113

So, You Bought a Baofeng Radio – Now what?

Channel Modifications & Review

First – Note that if you have already pulled in repeater settings through the prior section *DOWNLOADING REPEATERS FROM REPEATER BOOK*, you should not need to do any programming through the following process. These instructions do allow you to make some adjustments such as changing the name of the channel if needed.

In Chirp, the main Memory screen that will come up displays all 128 of the channels (0-127) in a spreadsheet-style layout. Each row represents a channel and each column is a setting. Not all settings can be done from this screen, however most of the needed settings can be. The next three screens are snippets of the Memory screen showing the columns and some examples. The first screen is a large snippet of the screen, followed by two closer views.

Loc	Frequency	Name	Tone Mode	Tone	ToneSql	DTCS Code	DTCS Rx Code	DTCS Pol	Cross Mode	Duplex	Offset	Mode	Power	Skip
41	162.550000	WX 1	(None)							off		FM	Low	S
42	162.400000	WX 2	(None)							off		FM	Low	S
43	162.475000	WX 3	(None)							off		FM	Low	S
44	162.425000	WX 4	(None)							off		FM	Low	S
45	162.450000	WX 5	(None)							off		FM	Low	S
46	162.500000	WX 6	(None)							off		FM	Low	S
47	162.525000	WX 7	(None)							off		FM	Low	S
48	146.790000	W4BCIV	Tone	146.2						-	0.600000	FM	High	
49	444.225000	W4BCIU	Tone	146.2						+	5.000000	FM	High	
50	442.550000	NI4-RV	Tone	100.0						+	5.000000	FM	High	
51	444.425000	NI4-TB	(None)							+	5.000000	FM	High	
52	442.825000	NI4-BT	Tone	100.0						+	5.000000	FM	High	
53	443.450000	NI4-HD	Tone	100.0						+	5.000000	FM	High	
54	442.950000	NI4-V1	(None)							+	5.000000	FM	High	
55	444.312500	NI4-V2	(None)							+	0.600000	FM	High	
56	145.430000	NI4-V3	Tone	100.0						+	5.000000	FM	High	
57	443.950000	NI4-LP	Tone	100.0						+	5.000000	FM	High	
58	146.520000	2MCALL	Tone	100.0						(None)		FM	High	
59	446.000000	70CALL	Tone	100.0						(None)		FM	High	
60	442.825000	S-ANDYT	Tone	110.9						+	5.000000	FM	High	
61	444.400000	S-APALA	Tone	94.8						+	5.000000	FM	High	
62	442.825000	S-BROOK	(None)							+	5.000000	FM	High	

Menu 1: The Menu list shows channels 0-127 and their current settings. Many of the settings here you can modify in this exact screen – Similar to a spreadsheet.

So, You Bought a Baofeng Radio – Now what?

The above screen is the main screen that you will be working with – however the image we have here is extremely difficult to read. For making it easier, on the next page we will provide a section with rows 48 thru 50 that are split into a left-side and a right-side of the screen in a much easier readable format with an explanation of each of these columns.

Settings	Loc	Frequency	Name	Tone Mode	Tone	ToneSql	DTCS Code
	48	146.790000	W4BCIV	Tone	146.2		
	49	444.225000	W4BCIU	Tone	146.2		
	50	442.550000	NI4-RV	Tone	100.0		

Menu 1-A: A closer view for better reading of the Menu-1 image at the top of this page. Columns include Frequency, Name, Tone type, Tone value, ToneSql, and DTCS Code value.

LOC – This is the channel number this row represents.

FREQUENCY – This is the frequency of the radio receiver. For Simplex this is the same channel for Transmit and Receive – For repeater channels this will be the frequency you are listening to.

Name – The friendly Name of the channel. This cannot be set in the MENU, this has to be set using CHIRP or software.

Tone Mode – Sets the Tone mode (None, TSQL, DTCS or Cross)

Tone – Sets the TONE Value if one is being used.

ToneSQL – Only used if TSQL is used in the above TONE MODE.

DTCS Code – Only used if DTCS is used in the above TONE MODE.

So, You Bought a Baofeng Radio – Now what?

DTCS Pol ◄	Cross Mode ◄	Duplex ◄	Offset ◄	Mode ◄	Power ◄	Skip
		-	0.600000	FM	High	
		+	5.000000	FM	High	
		+	5.000000	FM	Med	

Menu 1-B: A closer view for better reading of the Menu-1 image on the prior page. Columns include Frequency, Name, Tone type, Tone value, ToneSql, and DTCS Code value.

DTCS POL – This is the polarity of the DCS Code IF a DTCS code is used in the settings above.

Cross Mode – This is set if the radio uses more complex tones.

Duplex – This is set for the direction the transmit frequency is offset from the receive frequency. Values are +, -, Split, Off or None.
(TURN THIS TO OFF TO PREVENT TRANSMITTING ON A CHANNEL.)

Offset – This is the Offset value when communicating with a repeater. Usually this will be .600 for VHF, and 5.00 for UHF.

Mode – This is the mode of the radio. VHF/UHF Amateur usually FM. Other options are WFM (Wide FM), NFM (Narrow FM), AM and DV.
(The Baofeng will not support AM or DV)

Power – The power setting of the Radio. 8-Watt radios will have Low/Med/High as options. The 5-watt radios only support Low/High.

Skip – Sets the SKIP for the channel during scan mode. Options are None, S (Skip) or P (Priority). Set a channel to SKIP to prevent scanning mode for checking the channel. For instance – If you program the weather channels into your radio, you will want to skip them when scanning.

What about other settings?

So, You Bought a Baofeng Radio – Now what?

Yes – There are setting that cannot be set in this "Spreadsheet" mode. For those, you can right-click on the row and in the pop-up you can click on PROPERTIES and bring up the following pop-up to set the channel properties.

How do I find these settings for a repeater?

You really should not need to make setting changes through the Memory Spreadsheet process we just described if you followed the instructions for querying repeaters from the earlier section in this chapter for "DOWNLOADING REPEATERS FROM REPEATER BOOK".

So, You Bought a Baofeng Radio – Now what?

Advanced CHIRP Topics

In addition to what we have already covered, there are a few capabilities I would like to cover very briefly.

- POM Message – This is the POWER-ON Message for the radio that can be set through the SETTINGS → Other Settings menu.
- VHF & UHF Limits – Can be set through the SETTINGS → Other Settings menu.
- VHF TX and UHF TX can be disabled if desired or necessary through the SETTINGS → Other Settings menu.
- If you need to disable the keyboard, you can do so through SETTINGS → Work Mode Settings → Keypad lock. BUT – This will totally lock the keyboard from any changes, even from being able to change channels.

CHIRP is an excellent programming tool – And it is free. In comparison to the Baofeng's own software, I prefer CHIRP. It allows easy sharing of

Chapter 12: Hand programming the Baofeng

Of the chapters in this book – This is one of the most important chapters. If you find yourself in a situation that you ONLY have this book and your Baofeng – And no Internet, and no computer access for programming, then you need to understand how to program your radio by hand. This includes not only adding a frequency for SIMPLEX communication with a nearby radio operator, but also being able to program for a known repeater that would get you additional range capabilities.

When programming the Baofeng, this is ALWAYS done from the radio in FREQUENCY mode. Think of this as "Blank Worksheet" when in this mode. We will make all of the settings necessary for the radio to operate with our other operators, or with our repeater while in frequency mode. Once all of the settings are complete, then as our last programming step we will save the settings to a specific channel.

Note on the following instructions: Updates to the models and firmware of the Baofeng radios may alter these instructions between models and firmware releases. You should verify these procedures on your particular radio and If you do find differences make sure you have the radio manual available that has the exact sequences of steps or you copy those pages from your radio operators manual and keep that manual page or pages with this book for future reference. Those instructions will be on 1-2 pages.

First, we will set up the user radios that will be using the repeater. These radios will be set up the same for both operators. For the operator radios, we will be setting up both radios for a repeater with a repeater offset. Following are the settings for these:

So, You Bought a Baofeng Radio – Now what?

User Radio: (Set up to communicate with a Repeater)

Menu Option	Menu #	Setting	Comments
Frequency set to for receive*	Freq Mode	146.900	Receive frequency for user radios
TXP (Transmit Power)	2	Low	Setting to low power for testing
TDR (Dual Channel Monitor)	7	ON	Set so dual monitor is available (Ch A & B)
R-CTCS (Receive CTCSS)	11	250.3 Hz	Sets Receive Squelch Tone
T-CTCS (Transmit CTCSS)	13	250.3 Hz	Sets Transmit Squelch Tone
SFT-D (Shift Offset Direction)	25	Negative (-)	Shift Direction for Repeater use
Offset (Value of Freq Offset)	26	000.600	Shift amount for Repeater use

Follow these steps to configure the radios:

1. Set the radios to FREQUENCY mode. Do this on the UV-5R/BF-F8HP Style radios using the **[VFO/MR]** button, or on the UV-82 series radios by holding the **[MENU]** button while turning on the radio.

2. Press the **[A/B]** button or the **[EXIT/AB]** button (Depending on your radio) to choose the upper "A" channel of the radio. The radio can only be programmed when on the upper "A" channel. (UV-5R & BF-F8HP style radios)

3. For UV-5R/BF-F8HP radios press the **[BAND]** button on the UV-5R/BF-F8HP style radios to select the proper VHF or UHF band for programming. Note this is not needed on the UV-82 series of radios.

4. Enter the frequency **146.900** into the radio.

5. Press the **[MENU]** button and then "**002**" for menu #2, TXP (Transmit Power). For testing purposes, we are going to set our power to low which is 1-Watt. Make this setting by pressing the

So, You Bought a Baofeng Radio – Now what?

[MENU] button again to ENTER the menu change mode, and then the UP arrow until LOW appears on the display. Press [MENU] again to save the new value.

6. Press the [MENU] button and then "011" for menu #11, R-CTCS (Receive CTCSS). Make this setting by pressing the [MENU] button again to ENTER the menu change mode, and then the DOWN arrow until 250.3HZ appears on the display. Press [MENU] again to save the new value.

7. Press the [MENU] button and then "013" for menu #13, T-CTCS (Transmit CTCSS). Make this setting by pressing the [MENU] button again to ENTER the menu change mode, and then the DOWN arrow until 250.3HZ appears on the display. Press [MENU] again to save the new value.

8. Press the [MENU] button and then "025" for menu #25, SFT-D (Shift Direction). Make this setting by pressing the [MENU] button again to ENTER the menu change mode, and then the DOWN arrow until "-" appears on the display. Press [MENU] again to save the new value.

9. Press the [MENU] button and then "026" for menu #26, Offset (Value for Frequency Offset). Make this setting by pressing the [MENU] button again to ENTER the menu change mode, and then then enter "000 600" arrow until 000.600 appears on the display. Press [MENU] again to save the new value.

10. For the last step, we are going to save our new settings to a channel. Press the [MENU] button and then "027" for menu #27, MEM-CH (Memory Channel Set). Enter memory channel setting mode by pressing the [MENU] button again, and then to assign our settings to channel 125 enter "125" and press [MENU] again to save the radio settings to channel 125.

So, You Bought a Baofeng Radio – Now what?

REPEATER RECEIVING RADIO

Next, we are going to program our Repeater Receiving Radio. This radio will have a slightly different setting. For this radio, we are going to set it to receive our transmissions, and we will not need to set up a Shift Value or Offset values (Menu options 25 & 26) as we are using two radios and each radio will work from a separate frequency.

Repeater RECEIVING Radio: (Set up as our RECEIVE radio for our 2-Radio Repeater)

Menu Option	Menu #	Setting	Comments
Frequency set to for receive*	Freq Mode	146.900	Receive frequency for user radios
TXP (Transmit Power)	2	Low	Setting to low power for testing
TDR (Dual Channel Monitor)	7	ON	Set so dual monitor is available (Ch A & B)
R-CTCS (Receive CTCSS)	11	250.3 Hz	Sets Receive Squelch Tone
T-CTCS (Transmit CTCSS)	13	250.3 Hz	Sets Transmit Squelch Tone

Follow these steps to configure the radios:

1. Set the radios to FREQUENCY mode. Do this on the UV-5R/BF-F8HP Style radios using the **[VFO/MR]** button, or on the UV-82 series radios by holding the **[MENU]** button while turning on the radio.

2. Press the **[A/B]** button or the **[EXIT/AB]** button (Depending on your radio) to choose the upper "A" channel of the radio. The radio can only be programmed when on the upper "A" channel. (UV-5R & BF-F8HP style radios)

3. For UV-5R/BF-F8HP radios press the **[BAND]** button on the UV-5R/BF-F8HP style radios to select the proper VHF or UHF band for

So, You Bought a Baofeng Radio – Now what?

programming. Note this is not needed on the UV-82 series of radios.

4. Enter the frequency **146.900** into the radio.

5. Press the **[MENU]** button and then "**002**" for menu #2, TXP (Transmit Power). For testing purposes, we are going to set our power to low which is 1-Watt. Make this setting by pressing the **[MENU]** button again to ENTER the menu change mode, and then the **UP** arrow until **LOW** appears on the display. Press **[MENU]** again to save the new value.

6. Press the **[MENU]** button and then "**011**" for menu #11, R-CTCS (Receive CTCSS). Make this setting by pressing the **[MENU]** button again to ENTER the menu change mode, and then the **DOWN** arrow until **250.3HZ** appears on the display. Press **[MENU]** again to save the new value.

7. Press the **[MENU]** button and then "**013**" for menu #13, T-CTCS (Transmit CTCSS). Make this setting by pressing the **[MENU]** button again to ENTER the menu change mode, and then the **DOWN** arrow until **250.3HZ** appears on the display. Press **[MENU]** again to save the new value.

8. For the last step, we are going to save our new settings to a channel. Press the **[MENU]** button and then "**027**" for menu #27, MEM-CH (Memory Channel Set). Enter memory channel setting mode by pressing the **[MENU]** button again, and then to assign our settings to channel **126** enter "126" and press **[MENU]** again to save the radio settings to channel 126.

So, You Bought a Baofeng Radio – Now what?

REPEATER TRANSMITTING RADIO

Lastly, we are going to program our Repeater Transmitting Radio. This radio will have a slightly different setting from our prior radios. For this radio, we are going to set it to Transmit our transmissions, and we will not need to set up a Shift Value or Offset values (Menu options 25 & 26) as we are using two radios and each radio will work from a separate frequency. We are also going to have to make an adjustment to our VOX setting for automatic transmitting without having to use the PTT button.

Repeater TRANSMITTING Radio: (Set up as our RECEIVE radio for our 2-Radio Repeater)

Menu Option	Menu #	Setting	Comments
Frequency set to for receive*	Freq Mode	146.900	Receive frequency for user radios
TXP (Transmit Power)	2	Low	Setting to low power for testing
VOX (Voice Activated TX	4	10	Set to 10 to turn on – You may need to adjust this down
TDR (Dual Channel Monitor)	7	ON	Set so dual monitor is available (Ch A & B)
R-CTCS (Receive CTCSS)	11	250.3 Hz	Sets Receive Squelch Tone
T-CTCS (Transmit CTCSS)	13	250.3 Hz	Sets Transmit Squelch Tone

Follow these steps to configure the radios:

1. Set the radios to FREQUENCY mode. Do this on the UV-5R/BF-F8HP Style radios using the **[VFO/MR]** button, or on the UV-82 series radios by holding the **[MENU]** button while turning on the radio.

2. Press the **[A/B]** button or the **[EXIT/AB]** button (Depending on your radio) to choose the upper "A" channel of the radio. The radio can only be programmed when on the upper "A" channel. (UV-5R & BF-

So, You Bought a Baofeng Radio – Now what?

F8HP style radios)

3. For UV-5R/BF-F8HP radios press the **[BAND]** button on the UV-5R/BF-F8HP style radios to select the proper VHF or UHF band for programming. Note this is not needed on the UV-82 series of radios.

4. Enter the frequency **146.900** into the radio.

5. Press the **[MENU]** button and then "**002**" for menu #2, TXP (Transmit Power). For testing purposes, we are going to set our power to low which is 1-Watt. Make this setting by pressing the **[MENU]** button again to ENTER the menu change mode, and then the **UP** arrow until **LOW** appears on the display. Press **[MENU]** again to save the new value.

6. Press the **[MENU]** button and then "**004**" for menu #4, VOX (Voice Activated Transmit). We are going to set our sensitivity level to 10 requiring the strongest signal. Make this setting by pressing the **[MENU]** button again to ENTER the menu change mode, and then enter "**10**". Press **[MENU]** again to save the new value.

7. Press the **[MENU]** button and then "**011**" for menu #11, R-CTCS (Receive CTCSS). Make this setting by pressing the **[MENU]** button again to ENTER the menu change mode, and then the **DOWN** arrow until **250.3HZ** appears on the display. Press **[MENU]** again to save the new value.

8. Press the **[MENU]** button and then "**013**" for menu #13, T-CTCS (Transmit CTCSS). Make this setting by pressing the **[MENU]** button again to ENTER the menu change mode, and then the **DOWN** arrow until **250.3HZ** appears on the display. Press **[MENU]** again to save the new value.

So, You Bought a Baofeng Radio – Now what?

9. For the last step, we are going to save our new settings to a channel. Press the **[MENU]** button and then "**027**" for menu #27, MEM-CH (Memory Channel Set). Enter memory channel setting mode by pressing the **[MENU]** button again, and then to assign our settings to channel **127** enter "127" and press **[MENU]** again to save the radio settings to channel 127.

That's it! If you follow these steps you will be able to program the radio for SIMPLEX mode, or for use with a Repeater. There are additional settings that some repeaters may require such as R-DCS, T-DCS, or R-Tone (UV-82 Only). You may need to add these settings depending on the repeater you are connecting to.

So, You Bought a Baofeng Radio – Now what?

Repeater Planning if your offline?

What if the power is out, or the Internet is offline, and you need to program your radio for a nearby repeater? There are two methods I would recommend for getting lists of nearby repeaters.

1. **RadioReference.com**

    Using RadioReference.com you can look up radio frequencies and repeaters for amateur use by state and county. Search for your specific location and look for the tab "Amateur Radio".

2. **CHIRP**

    Chirp is actually my preferred method. With Chirp, you can either export the current repeaters and settings in your radio to a CSV file for printing, or you can use the RADIO → Query Data Source → Repeater Book and pull a list of repeaters for 2-Meters or 70-CM frequencies by location.

If you have purchased the radio for a "Just-In-Case" need, we are going to STRONGLY recommend that you make a list of repeaters in your area, or an area that you may be relocating to in the event of an emergency. Print this list, and then store it with your radio and this book so that if you do need to use the radio, you have a roadmap of frequencies to use. Remember that without a proper tone and off-shift settings for hitting a repeater, you will be wandering blind trying to get the radio to work when you may absolutely need it.

So, You Bought a Baofeng Radio – Now what?

Chapter 13: Lingo & Talking on the Radio

One of the most difficult things for too many new Hams is actually getting onto the radio and talking. You don't know anyone and being new you may not know what to say. Well we will cover just a few quick basics here so that you understand some of the lingo and the protocol.

First – You are at a Tech level probably. I am assuming this because this is a more entry level book to get new folks started with the Baofeng Radio. What is some of the lingo that a new Tech needs to get used to using when talking on the radio. In chapter nine we covered finding nets, and then in chapters ten and eleven we covered programming your radio. So, at this point you should be able to jump on and start talking. You are going to need to look for a repeater first to increase your chances of reaching someone. It would be better to find a busy repeater or a repeater network, and for that you will have to do some research using some of the repeater site directories we mentioned in chapters seven and eight.

Initiating Communication

Once you find your channel you want to talk on, one good way to start is simply address the channel – perhaps just look for a radio check or call any station. To do so, do the following:

1. Announce your call sign. "KX4ABC, this is Rodney is Tampa".

2. Then what are your doing. Here are some examples:
    - "Calling any station" if you are making a general call for anyone.

    "Calling for a Radio check" if you just want to confirm you are being heard. Responders may come back just to give a quick confirmation not intending to converse – but either way you have confirmed your radio works.

So, You Bought a Baofeng Radio – Now what?

3. If you are trying to reach someone specific, "Calling KX4XYZ" and wait for a response.

Those are the simple steps to initiate a call. Not difficult at all. With the radio check expect that if someone hears you, they will give you a quick signal report. They will let you know if you are coming in clear, or with static. If you are calling in on the fringe of the repeater, you may be able to trigger the repeater but not be heard. In this case either someone will respond back to the station not being heard or will ignore your call.

Ending a Communication

When you complete a communication, usually it is done one of two ways. Finish talking to the other party, and then at the end simply say "73". The term 73 actually comes from a short-hand code developed in 1879 by Walter P. Phillips for Morse code operators to end their communication. Years later this, along with a few other shortcuts, were picked up by ham radio operators. 73 is still frequently used today to indicate the end of your communication. It doesn't mean you are signing off the radio, but just that you are ending communication with the person you were speaking with. The exact wording you would use is "This is KX4ABC, 73". Remember you have to speak your call sign at the end of your communication for identification purposes.

Another simple method to indicate you are going to be no longer talking but are there on the radio listening is to end your conversation with either "This is KX4ABC, Monitoring" or "This is KX4ABC standing by". These two ways of ending a communication are also used if you have tried initiating a conversation with no response. So say you log onto a repeater to talk but reach nobody. Here is what that would look like: "KX4ABC, This is Rodney from Tampa. Calling any station."

So, You Bought a Baofeng Radio – Now what?

Then pause and wait for a response. If nobody comes back, end your call out with "This is KX4ABC Standing By" (Or Monitoring).

Call Identification

In Ham radio, the FCC requires that you identify yourself with your call sign every ten minutes during a conversation. You need to identify yourself at the beginning, and the end, and every ten minutes. If you are in a long conversation that exceeds ten minutes, then at the ten-minute (Or so) mark you would state "This is KX4ABC for identification".

CQ, CQ, CQ

Another phrase you have likely heard on TV or in a movie, or on ham radio, is "CQ CQ CQ". This is short for calling any station. Now on VHF/UHF with repeaters this is not generally used because this is normally used when looking for a long-distance contact or an overseas contact. Since we are using local FM analog repeaters we don't use CQ, CQ, CQ.

CB Radio Lingo – Or, "Hey – They 80's just called and they want their Lingo Back, Good Buddy!"

Don't. Just don't. Except for a few short codes, on FM Ham Radio we don't try to leverage any type of codes or jargon. Saying "10-4" might be acceptable if is part of your everyday vocabulary, but on Ham radio in general we speak in clear simple everyday speech.

In Ham radio it is forbidden to use any special codes or encryption to hide the meanings of your conversation. Terms like "73" are established and allowed, but don't be using and special coding over the radio or you will be attracting attention and breaking the rules.

Ham operators in general are a very friendly bunch. Many are reaching out and just trying to make contacts and conversation. In Chapter 14

So, You Bought a Baofeng Radio – Now what?

coming up next we are going to touch on finding a club. Getting plugged in to other Ham operators will help acclimate you to getting to talk on the radio. If you are getting into Ham Radio from the Prepper community – then likewise look for a local Prepper or Readiness group. We will touch on that also in chapter 14.

Chapter 14: Finding a Club

Participating in a Ham club will put you in touch with a group of Ham operators also very interested in the hobby. Through club participation I myself have made contacts that have allowed me to meet and work with local Emergency teams, learn emergency radio, and have allowed me to get into interesting projects such as the build of our emergency radio van – a project that will provide our local club with a radio-ready van that can be dispatched within Florida for emergency needs.

Finding a club is as easy as visiting the ARRL's website and searching for clubs based on your state and local information:
http://www.arrl.org/find-a-club

There are clubs that meet in person and clubs that meet virtually. My own club has shifted to virtual meetings recently due to the 2020 COVID outbreak. We still meet in person for small events such as our Ham testing sessions, and special projects or events in small teams such as our Van build project and field events.

Clubs frequently participate in nationally sponsored events such as Field Day, Emergency Planning exercises, Fox Hunts, seminars and educational events. We work with school students and scouts to show the value and importance of Ham communication, and we work with our county and state officials in Ham Radio emergency planning.

From a Preppers Perspective
Today many folks are getting into Ham Radio from the perspective of a Prepper or emergency planner wanting to be in communication in the case of an emergency, natural disaster, or a worst-case SHTF scenario which could impact the power grid and normal communications. In the

So, You Bought a Baofeng Radio – Now what?

classes I teach about half of those participating also participate in some kind of readiness planning. Living myself in the Tampa Bay area of Florida, we are constantly watchful of Hurricanes. In 2018 hurricane Michael left tens of thousands in the panhandle of Florida without power and communications for weeks with Cell service being solidly knocked out for over two weeks.

Finding a Prep group is not as easy – and you do need to be more selective as well. Though most groups are just fine, there are some who are fringe groups that you don't find in Ham clubs. A good starting source would be to look at a national group called PrepperNet (https://www.preppernet.com/) that focuses on educational topics ranging from Ham Radio to Food preservation, Operational Awareness, bugging out, and topics related to survival. I participate in one of these groups where we focus heavily on skills you would need and relationship building that would help families for survival, communicating and homesteading.

Facebook and Meetup are two other good sources for finding local groups. Many groups do not actually meet in person, but for those that do there will usually be a sub-group of Ham operators and radio guys that will be happy to discuss radio with you and get you acclimated to local events and activities. My local group is planning this fall to hold Ham classes and emergency radio discussions that extend to CB, GMRS and MURS radios as alternatives for non-Hams.

So, You Bought a Baofeng Radio – Now what?

Chapter 15: Basics About Coax & Your Baofeng

For the most part you will be using your Baofeng radio with an antenna attached directly on the radio. But what about if you need better performance from the radio, or you need to connect to an antenna mounted to your car for better operation?

In these cases, you can connect an external antenna to the radio that will be connected through a segment of Coaxial cable. In Ham radio we generally use the RG-58 type of cable which is a 50-ohm cable. This happens to be the same cable used in most CB Radio installations, so if in a pinch you may be able to pick up cable extensions from a local Truck stop.

You probably won't be using Coax unless you decide to use an external antenna with your radio.

SMA Female to BNC Female Adapter for the Baofeng Radio
https://www.amazon.com/DHT-LLC-AD075-Convert-Adapter/dp/B00CVQK466/ref=sr_1_11?dchild=1&keywords=SMA+to+BNC+adapter&qid=1598476663&sr=8-11

So, You Bought a Baofeng Radio – Now what?

DHT Handheld Antenna Cable for Wouxon/Baofeng SMA Female to UHF SO-239 Female Connector

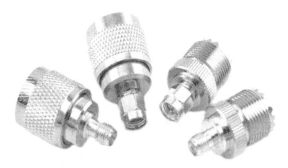

SMA-UHF RF Connectors Kit

BNC to UHF – 4 Types RF Connector Kit
https://www.amazon.com/Connector-Coaxial-Adapter-Adapters-Straight/dp/B074R7TP1R/ref=sr_1_4?crid=1QNRKOS6Q2K65&dchild=1&keywords=so239+to+bnc+adapter&qid=1598472159&sprefix=SO239+to+BNC%2Caps%2C174&sr=8-4

So, You Bought a Baofeng Radio – Now what?

PL259 UHF Mail Plug to BNC Female Right angle 90 Degree adapter
https://www.amazon.com/exgoofit-Connector-degrees-adapter-Adapter/dp/B07CQGC2K1/ref=sr_1_16?crid=1QNRKOS6Q2K65&dchild=1&keywords=so239+to+bnc+adapter&qid=1598472325&sprefix=SO239+to+BNC%2Caps%2C174&sr=8-16

RG-58 15 Meters (49 Ft) UHF SO-239 Female Connectors
https://www.amazon.com/PL-259-Antenna-Coaxial-Connectors-Extension/dp/B078V56SMV/ref=sr_1_2_sspa?dchild=1&keywords=RG-58+Coax&qid=1598476422&sr=8-2-spons&psc=1&spLa=ZW5jcnlwdGVkUXVhbGlmaWVyPUExVFo4MEZPUzcyR0U3JmVuY3J5cHRlZElkPUEwODc4MDYzMlFPOVNBV01KN1ZFQyZlbmNyeXB0ZWRBZElkPUEwNjgzMTE1MUc1OVcyTk9CU0U0JGNCZ3aWRnZXROYW1lPXNwX2F0ZiZhY3Rpb249Y2xpY2tSZWRpcmVjdCZkb05vdExvZ0NsaWNrPXRydWU=

So, You Bought a Baofeng Radio – Now what?

3 Ft RG-58 Jumper Cable, SMA Female to UHF SO-239 Female

https://www.amazon.com/Handheld-Antenna-Wouxun-Baofeng-Quasheng/dp/B00HX18TJI/ref=sr_1_23?dchild=1&keywords=RG-58+Coax&qid=1598476176&sr=8-23

Chapter 16: Are you going to Use that Radio?

This is an important chapter. I have spoken to so many folks – mostly in the Prepping community – that have purchased their radios and they toss them into a box or into a faraday cage. They assume they will be able to use them when they need them, but by not using them they never learn *How* to use them.

What should you do? Better to have the radio than not when you need it, and learn how to use it when you might be under stress at some point and can take a few hours then to learn? Nope.

First of all, you need to get the radio programmed. Preferably using CHIRP, but by hand if you can't use a computer or don't have the programming cable. You should also print a list of repeaters throughout your area and store that with the radio. If an event occurs which disrupts communication and the Internet, you will not be able to look these up. You can download these using Chirp and print them out to have on file.

You also have this book – Keep it – and make sure you have your original radio manual handy and nearby. This book could be incredibly valuable with the references to frequencies we have in the back as well as programming instructions. Remember if the Internet goes down you can still program the radio with Chirp.

If you have not purchased the PC-03 cable with the FTDI chip in it – get it. Clone cables that need the older Prolific 3.2.0.0 driver will be useless if the Internet is down and you cannot get the drivers.

When you write up and print references to frequencies in your area, don't forget to look for repeaters that may be in the alternate area that

So, You Bought a Baofeng Radio – Now what?

you may leave for. If you plan to stay with friends or relatives a distance from where you are, consider creating printed references to the repeaters in those areas and for areas along the way. I have family myself a couple of hours north and have the repeaters notated for the entire way up. The types of events I plan for are really storm and hurricane related, so an evacuation out of Tampa would take me through about a half dozen different radio repeaters that I could use to contact other Hams for evacuation and road conditions. My real concern would be less for the evacuation out of the area and more for the trip back in where Cellular communications could be down.

So, in summary, the worst thing you could do with that new radio is to just toss it into a box. Get your license, and barring that get it programmed. We have provided you enough information in this book to get you to using CHIRP and to get the radio programmed. Getting local repeater channels on it, and even the simplex frequencies as well as *GMRS and weather frequencies could help you in an emergency.

*It is illegal to transmit on the GMRS and Weather frequencies, but these frequencies could be added to the radio to receive transmissions only.

Things to do following purchase of your radio:

- Get the proper FTDI Chip Radio Cable
- Charge the radio
- Install CHIRP
- Program the radio with essential simplex channels
- Program the radio via CHIRP with repeater channels.
- If you're a TECH licensed ham, confirm the radio works by checking into some local nets.
- Upgrade the Antenna and purchase at least 1 spare battery.

Chapter 17: Advanced Topics

SCANNING

The Baofeng radio does have a scanning capability. It can scan through frequencies while in frequency mode, or scan through the 127 channels that are programmed into the radio.

But make no mistakes – The Baofeng radio is NOT a good scanner. When in channel mode and scanning through a limited number of pre-programmed channels it is adequate, since there are not that many to really scan through. You can limit which channels get scanned while in channel mode through software programming such as with CHIRP and setting various channels to be skipped. You won't want to scan through the weather channels for instance, because the radio will nearly always stop on those channels as they constantly broadcast.

When in Frequency mode you can speed up scanning by increasing the scanning step size (MENU Option #2) but realize by doing so you will also be skipping by frequencies that could be carrying traffic.

The Baofeng Radio really is not a good scanner for frequency scanning. Also – If you are intending to use if for scanning public service frequencies, realize that many municipalities are now moving digital radio, sometimes encrypted, and often to Trunking radio systems. These systems allow greater use of the frequencies and the bandwidth, but this also makes monitoring from a simple radio like the Baofeng impossible. If you are looking for a scanning function you would be better with a scanner designed specifically for this purpose.

To Scan on the Baofeng radio, simply place the radio in either Frequency or Channel mode, then press the [*SCAN] button until the radio begins

So, You Bought a Baofeng Radio – Now what?

scanning. You can control how the radio reacts to a found transmission through the [MENU] "18" option with three separate modes – Time, Carrier and Search operation.

- TO Mode - Time Operation: When a signal is found, scanning stops until a pre-defined timeout, then resumes scanning.
- CO Mode – Carrier Mode: The scanning stops when a signal is detected and then resumes after a pre-defined time with no signal.
- SE Mode – Search Operation: Scanning stops when a signal is detected, and then starts back only when the [*Scan] button is pressed.

DUAL WATCH

Dual watch on ham radios is accomplished either by having the radio able to rapidly scan between two frequencies or channels, or by the radio having two receivers built into it.

The later method is a more expensive method found in more expensive radio systems. The Baofeng radio actually uses the first and less expensive method. What this means is that when a signal is heard, the radio locks onto the frequency or channel it is receiving the signal on until that received signal stops. Then it will again monitor both frequencies again alternating rapidly between both.

A radio with two receivers in it can actually allow you to hear both transmissions at the same time though this is often unnecessary.

CHOOSING A CHANNEL

This is simple enough, but let's just take a second to cover channel selection with the Baofeng Radio.

First – To use channels, you must have the radio in "Channel Mode". For the UV-5R/BF-F8HP style of radio, this is a matter of selecting Channel

So, You Bought a Baofeng Radio – Now what?

mode with the [VFO/MR] button. For the UV-82 series of radio this is done by holding the [MENU] button down WHILE turning on the radio.

When in Channel mode, you can either key Up and Down channels using the UP and DOWN buttons, or you can key in the channel directly through the keypad. All channels below 100 must be proceeded with a "0", so to key in channel 50 key in "050". To key in channel 125 key in "125".

CHANGING POWER LEVEL

Power levels can be programmed into the radio along with the memory channel. This means that if you have a channel programmed it will always load with the power level you designate when programming the channel. For all 5-watt radio our power options will be Low and High for 1 watt at low power, and 4/5 watts for High settings. (The 5-watt radios actually work at 4-watts for UHF, 5-Watts for VHF).

For all 7/8 watt radios, our settings are 1-4/5-7/8 watts. (The higher watt radios also operate at a different max output for UHF than for VHF)

While operating the radio, you can also alter the power levels by pressing and holding the [0] key on the keypad which will update the screen display between L-H for the lower watt radios, and L-M-H for higher watt radios.

TONE SCANNING

Tone Scanning is an important concept if you are looking to connect in with a Repeater but you don't know what the PL-Tone/CTCSS Tone is that the repeater is using. The scanning process can be done in either

So, You Bought a Baofeng Radio – Now what?

Frequency/VFO mode or in Channel/MR Mode, but the tone can only be saved while in VFO/Frequency mode through Menu options 10/11.

To scan for a CTCSS tone follow these steps:

1. Either go to a frequency in FREQUENCY mode or go to an already defined channel that you want to scan.
2. Enter MENU mode by pressing [MENU]
3. Go to Menu option 11 – R-CTCSS
4. Press [MENU]
5. If R-CTCSS is set to "Off" then hit the UP or DN button to set a value, then hit [MENU] to save the value. Note: This is not the correct CTCSS used by the repeater frequency, but the radio must be set to a tone before scanning can be done.
6. If the radio R-CTCSS is now set to a tone, even if not right, then hit [MENU] as if to change the CTCSS value. While in selection mode, not hit the [*SCAN] button. A Flashing "CT" will display in the LED Display indicating the radio is in scan mode.
7. Once a transmission comes onto the radio and the repeater is functioning, the radio will begin scanning and rotating between CTCSS codes until it lands on the code being received in the transmission. At that point it will stop and you now know the code being used with the repeater.

Now that you have your tone you can use this value to program your radio for a specific repeater. In addition to scanning for the CTCSS tone, you can also scan for the DCS tone if being used on the repeater. To do so, perform the same steps above however instead of using Menu Option 11 for R-CTCSS in step #3, use Option 10 for R-DCS.

Some Baofeng Radios, particularly the UV-82 series of radios, also support TONE-BURST for amateur radio systems. This can be set through

So, You Bought a Baofeng Radio – Now what?

Menu option 41, only available in the UV-82 series and can be set to a limited number of values: 1000 Hz, 1450 Hz, 1750 Hz or 2100 Hz.

VFO LOCKOUT

The VFO Lockout function on the Baofeng Radios is used to prevent field programming. The VFO lockout function is set through software and cannot be set from the radio menu. This is also an option available only on some model radios, and specifically some models of the UV-82 radios such as the Commercial UV-82C from Baofeng Tech.

The ability to lock out the keypad is required for use as a Commercial Radio (The UV-82C). For details about if VFO Lockout mode is available on your specific radio, you need to check the manual for your radio.

POM MESSAGE

The POM Message is the POWER-ON message that is displayed on the radio when you first turn it on. For my own radio, I set the first line to my last name and the second line to my call sign. This makes it easier to get my radio back if misplaced or lost. Of course, I still tag the radio the old fashion way with a sticker on the back of the radio.

TX AND RX FREQUENCIES WITH OFFSETS

When programming channels on the Baofeng radio which will have different Transmit and Receive frequencies, such as when setting up a channel for a repeater, you will not actually program both TX and RX frequencies. Instead, you will program into the radio the RX frequency, the OFFSET amount, and the OFFSET direction. By programming all three of these into the radio you will be giving the radio all it needs.

So, You Bought a Baofeng Radio – Now what?

For instance, if you enter a TX frequency into the radio of 145.000, and a POSITIVE (+) offset with a value of 000.600, then the transmit frequency will be 145.600.

ABOUT RADIO FIRMWARE
The Baofeng Radios come with a hard-programmed firmware and cannot be updated. Many newer style radios from other manufacturers come with firmware that can be flashed to provide the radio with updates that may include bug fixes or new features. For instance, I have a BTECH DMR-6x2 Radio which I purchased shortly after release. It is currently my daily radio due to many advanced capabilities and works both Digital and Analog communication. As I have updated firmware over time, I have received new capabilities with the radio.

The older and simpler Baofeng radios do not have this ability – But this does not detract from the strong capabilities that these radios do have. But if you do run into a problem that is associated with the firmware version of your radio, the only way to get around it will be through the purchase of a new radio.

Chapter 18: Non-Ham Communications

If you have your Baofeng radio, I would expect that you have it with the plans to obtain your Ham Radio license and get on the air. There really is no better radio option outside of Ham Radio to be able to get distance communications in the event of cell and telephone systems going down.

With your Tech License, a Baofeng handheld, and access to a repeater network you have a range of dozens to hundreds of miles. For Simplex, you are limited to between 2 and 5 miles generally but add a 50-watt mobile radio into your capabilities along with a good external antenna mounted high and you will extend to between 10- and 40-miles range for talking (Through a repeater). But – That is YOUR capability with the right equipment and the right equipment. What about others in your family?

The next two options I would recommend are GMRS and Citizens Band (CB) radio systems. Each of these will also have limited ranges. GMRS radios are the radios frequently sold at Walmart or the Sporting Good stores with an advertised range of 56 miles (That is a marketing Joke). Where that 56 miles actually comes from is you standing at the peak of a mountain and a friend 56 miles away at the top of another mountain, During a clear winter day, no rain, no clouds.

BTech
GMRS-50X1 50-Watt Radio

BTECH
GMRS-V1 Radio, 2+ Watts

So, You Bought a Baofeng Radio – Now what?

In reality GMRS has a range of 2-3 miles, just like your Baofeng. The stubby antenna on the radio will limit that, but you can purchase 5-watt GMRS Handhelds that allow external antennas to be used. Raise an antenna, and the radio will have similar performance to your Baofeng. There are also 50-watt mobile GMRS radios available from companies such as Baofeng Tech and Midland that will offer you a portable mobile or table-top style radio solution with and externally raised antenna. These radio systems are popular in rural farm or ranch environments. Licensing is minimal not requiring a test and lasting for ten years.

As for CB Radio – CB Radio has been around since the 70's when they took off like hotcakes in cars and trucks on the road. Even today most long-distance and independent truckers still can be found with CB Radios in their vehicles. Today they are used less for primary on-the-road communication thanks to newer communications tools including Cellular telephones, Voice-Over-IP radio (Such as Zello), and even Ham radio. But where they still shine are in the "Last Mile" trips allowing truckers to communicate with dispatchers at delivery and pickup depots, talking with other drivers in places such as truck and rest stops, and – emergency communication.

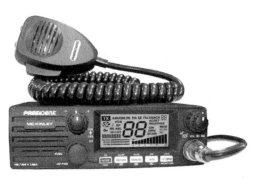

President McKinley 40-Channel SSB AM Radio

So, You Bought a Baofeng Radio – Now what?

So why would you want to fool with either GMRS or CB Radio as a part of your emergency communications? You have Ham Radio – Either your Baofeng Handheld, or a 50-watt Mobile Radio (Maybe a Baofeng there to?).

Good emergency communications planners take into account multiple ways of communicating. My ham radio club is currently in the process of designing and building out an older van as a Mobile Amateur Radio Vehicle (MARV) for use during a hurricane or disaster, or for deployment to a remote location after a disaster. (Such as after the 2018 Hurricane Michael disaster which left communications shattered for weeks). I have been on the team designing the specifications for the vehicle – What and how many radios, power, off-grid power, cooling, etc. Two of the additions that we are making will be a single CB Radio, 40-Channel SSB model, plus a single 50-watt GMRS Radio which can scan and monitor GMRS Channels. These two radio systems will not be our primary radios but will be secondary communications radios for use to monitor short-distance communications. With our antennas up on a 30' portable tower, or range for communications will be 10 miles plus the range of the other radio operators.

For your own planning you might consider something like this as well. A Ham operator with a Tech license able to access a repeater network, or a General license operator with a HF Radio can easily be a central "Belly Button" able to relay emergency messages. I strongly consider both GMRS and CB Radio key communication tools and neither should be overlooked.

So – How far will this get you?

If you and a family member BOTH have a CB or GMRS Radio system set up, with a good external antenna up about fifteen or twenty feet, then

So, You Bought a Baofeng Radio – Now what?

you could likely get somewhere between 10 and 20 miles between you both, but this is heavily dependent on so many factors just like with Ham radio.

- Terrain between you
- Mountains, Hills
- Quality of your Antenna (Higher and clearer is best)
- Number of Trees, leaves between you where you are talking to
- Number of cars
- Number of buildings
- Radio Power
- Etc….

In other words – Just like VHF/UHF Ham Radio in Simplex mode, lots of things can cut into your range. Between the two radio types, I would feel more comfortable with 50-watt GMRS Radios between two locations having the best signal.

But when relying on GMRS or CB to reach any random person, I would expect better from CB based on the number of CB Radios still in use by truckers and recreationalists. Most folks with GMRS Radios will have the smaller 1-5 watt bubble-pack radios with the permanent stubby antenna that will have a range of a half mile to maybe two miles.

Chapter 19: Getting Licensed & Legal

I will have three primary types of individuals who are purchasing this book.

1. A Readiness/Prepper type of individual looking for a good reference to go with their radio which they have purchased as a part of their Preps and really only have an interest in the radio for an emergency.

2. The Ham Radio student who has purchased the radio to use in learning Ham radio and while studying for their Ham Radio test.

3. The new Ham Radio Tech-Level Licensed operator still learning ham radio and you have a Baofeng learning to use it.

4. The existing Ham License holder, Tech Level or above looking for a Baofeng Book resource.

I am going to address each type of person separately.

1- Readiness/Prepper Radio User
So often I will hear a prepper say – "Well if SHTF a license isn't needed anyway", or "I just need the radio just in case".

But do you really know how to use the radio, and understand Ham Radio theory? I have covered I hope enough in this book to help you with the radio and to get it up and running. But operating the radio is only a small part of using it. You will learn so much through getting the license, and then that will open the door in using the radio legally. I am talking about

So, You Bought a Baofeng Radio – Now what?

participating in nets, conversing on the radio, and actually applying what you know to practical use.

Many people get into Ham radio to talk – to open doors to new areas, and to converse with others a long way away. Though the Internet and apps like Zello make getting your license or "Ticket" less necessary to do so, the use of Ham Radios still offers a lot of opportunity to plug into this community.

Many others, like myself, have gotten into Ham Radio for the "Emergency Capabilities" aspect of it. I have become fully licensed with my AE level license, plugged into clubs, and have volunteered to work with our local ARES Emergency Radio group which has extended my education far beyond just the license. I have learned so much more through group participation and exercise participation than I would have with just getting licensed – and worlds more than having just gotten a radio and set it aside.

2- The Ham Radio Student

For you – Great! You have decided to jump into Ham Radio. The Baofeng Radio is a great, low cost first radio. It will give you the tool you need to get started in radio.

At first you need to review what this book has brought to you and get the radio programmed and running. Look for clubs and repeater nets. Get the radio set up with CHIRP and get the repeaters programmed into it. Remember until you get your license you are not legally able to talk on the radio, but you can learn a lot through listening. If you are in an area that is prone to disasters, many of the radio nets will focus on those events when they occur. So far in the 2020 season the local repeater net covering 10 counties and 9,000 square miles has stood up twice with

So, You Bought a Baofeng Radio – Now what?

weather and planning news and announcements. I have passed to my students to tune in and listen and see what the radio is used for.

At this stage, you need to keep spending time studying and learning the material. If you have not already purchased a study book, you need to decide on your best method of study.

If you self-pace yourself and self-study, consider looking at the Gordon West Technician study book. This book divides the material into twenty categories based on topics and gives all of the exact questions and answers that the test will have. The book also goes into a good summary detail of the answers. This is the book I actually use in my classes as it has a great method of grouping questions.

The test consists of 35 questions from a pool of 423 questions. Of the questions, many are simply memorization questions – Just the types of questions that have to be memorized. Of the others, many are formula and math based, while others do require an understanding of the concepts which are explained in the GW Book. The ARRL also produces a good college-quality textbook that covers the material, though not as focused as the Gordon West books. Though they make you dig some, they do go into greater depths of explanation.

3- For the new Tech-Level License Holder

Once you get your Tech License, then the next steps you should be planning to move forward with for Ham Radio would be to consider purchasing a Mobile radio – 50 Watts of power that could be used either in your vehicle or at home. Connect a good external antenna and get busy talking – Just checking into some nets and learn the nets in the area.

So, You Bought a Baofeng Radio – Now what?

Realize that your Tech license has opened some doors for you. Distance communication for VHF/UHF is highly dependent on Repeater networks. The next level of consideration will go far beyond the scope of this book; however, I would recommend moving towards a General license. With this level of license, you can get into HF Communications – Lower frequencies, longer wavelengths that have the ability to bounce through the atmosphere and thereby extending your range to hundreds or thousands of miles without a repeater network.

If you are planning for emergency radio operation, loss of a repeater or repeater network, or an Internet backbone linking the repeaters together would reduce your distance and capabilities. For this – The General Operator's license along with HF Radio would be your only good method of long-distance communication. Short-range radio would still be available – Short-range Line-of-sight Ham VHF/UHF, GMRS and CB Radio would be working, but generally for distances under 15-20 miles. For anything else you would need to be able to work HF Radio (General License).

4- Existing Ham License Holder

For the experienced Ham license holder at any level, I really have no advice for you. By "Experienced" I assume you know Ham at whichever level you are at and have likely purchased my book as a reference to the radio. For you I thank you, and wish you good air!

So, You Bought a Baofeng Radio – Now what?

Resources:

- Book - Gordon West 2018-2022 Technician Class (Yellow Book)
- Book – The ARRL Ham Radio License Manual (Red Book)
- Book – So you Bought A Baofeng?
  http://www.SoYouBoughtABaofeng.com/

- Classes – ARRL Website
  https://www.arl.org/find-an-amateur-radio-license-class/
- Gordon West Radio School
  http://www.GordonWestRadioSchool.com/

- Practice – Ham Radio Prep
  http://www.HamRadioPrep.com/
- Ham Exam
  http://www.HamExam.org/
- Ham Study
  http://www.HamStudy.org/

So, You Bought a Baofeng Radio – Now what?

So, You Bought a Baofeng Radio – Now what?

Chapter 20: Some recommended Accessories

Now that you have your Baofeng, and hopefully either now or soon you will have your license, what else do you need?

Adding accessories to your radio will make it more convenient to use. You don't need any of course, and if you are still in the process of getting your license you probably don't need any. But I want to take some time here to cover various accessories that you can add to your radio to make it easier and more reliable to operate.

PROGRAMMING CABLE

For the programming of your radio I put this accessory at the top of your list. Programming gives you the ability to back up the current configuration of the radio and add/remove channels quickly. This is important if you are going to be going into another area and need the radio set up for that location.

The ONLY cable I would recommend is the PC-03 cable from Baofeng Tech (https://baofengtech.com/Programming-cable). This cable has the FTDI chip in it for interfacing with your computer and does not require special drivers. There are clone cables you can purchase and save a few dollars, but you will have to hunt down the drivers and if your computer operating system updates (When) you will have to repeat.

Baofeng Tech PC-03 Programming Cable

So, You Bought a Baofeng Radio – Now what?

ANTENNA CONNECTOR ADAPTERS

This one is absolutely and totally optional, but I have done this to make antenna swapping quick and easier. You can purchase a Female SMA to BNC Female Converter. You attach this onto the Baofeng radio after removing the antenna, and now you can attach any antenna or adapter that uses a BNC connector quickly and without having to untwist the SMA accessory or worry about cross-threading. I have a PL-239 to BNC adapter on my car Mag-Mount antenna and I also purchase now all of my antennas with BNC Connectors.

SMA Female to BNC Radio Antenna mount adapter

SMA to SO-239 Adapter

This adapter is needed to allow you to connect your Baofeng radio to an external Mag-Mount or Car-mounted antenna. If you plan to use the radio in a car then you will absolutely need a car mounted antenna, and you will then need one of these. Notice in the picture below there are two types – one with two connectors with a short cable between them, and one as a one-piece adapter.

So, You Bought a Baofeng Radio – Now what?

ANTENNAS

Next on your list you need to update the antenna that comes with the Baofeng. The short antenna is at best adequate, where a better-quality antenna will make it easier to pull in signals and to be heard on the radio. With antennas you will find a lot of different options. Here are some recommendations:

- Nagoya NA-771 – From Baofeng Tech or Amazon
- Signal Stuff "Signal Stick" – https://www.signalstuff.com/
- Abbree 42.5" Folding antenna – Find on Amazon.com

Be sure when checking for an antenna you get the right base mount. If you stick with the standard SMA Screw-on antenna mount (Without the SMA-to-BNC Adapter) make sure all of your antennas are SMA. Keep in mind – Even if you get the adapter, with an SMA antenna you can always remove the adapter and use the SMA connection.

You should also consider a good Mag-Mount antenna or maybe a car-mount/trunk-mount antenna if you're going to use the radio in the car or if using it outside. Getting the antenna where signals are transmitted and received away from blockages will greatly improve reception and chances of being heard by others. Baofeng Tech has some Nagoya antennas listed on their website as well as Amazon. If searching on Amazon, make sure the antenna you are looking at is a UHF/VHF Ham Radio antenna and not a CB antenna. Cost should be from just under $30 and up depending upon quality.

We listed other styles of Antennas in our chapter dedicated to them so please make sure to review that chapter.

So, You Bought a Baofeng Radio – Now what?

Antenna Poles
Depending on if you decide to erect an outdoor antenna or not, if you do then a very easy and simple solution is the Paint-stick solution I covered in the chapter for Antennas. A simple solution with some parts from Home Depot. Check out our Accessory links in Appendix I for a specific reference to one such pole from Home Depot.

Pie Pans or Pizza Pan
Pie Pans you ask? If you are going to try a Mag Mount antenna as an external antenna solution, adding a metal pie pan at the base does two things. First it adds stability to the antenna. Second – the metal pan will act as a ground plane for the antenna and give you better antenna performance. A simple and cheap solution.

Hand Microphones
A good hand microphone will benefit you tremendously in the car or if you are using the radio where you may wish to protect the radio – such as if hiking or in the rain. I have tried several microphones and have had a bit of trouble with getting the microphones to seat properly in the Kenwood plug on the radio. As of this date, the best microphone that I have purchased for any of my Baofeng radios is from Klein Electronics which specializes in two-way radio microphones.
(https://www.kleinelectronics.com/)

Baofeng's traditional hand microphone is only one of many available Mic styles. Using a hand microphone in the car is much more convenient and safe.

So, You Bought a Baofeng Radio – Now what?

Ear & Boom Microphones
An ear or Boom microphone will allow you to monitor the radio quietly. Your Baofeng came with an inexpensive ear microphone that works fine for this purpose. I carry a boom microphone which offers better microphone quality – but either work fine for monitoring.

Bluetooth Microphones
These are actually relatively new. With a Bluetooth microphone, you have a cordless hand microphone that you can carry and use to speak with. It connects via a small plug adapter that plugs directly into the accessory port on the side of the radio. One big advantage of this is not needing to keep the radio out – So you can protect the radio in bad weather. But – The manufacturer warns of possible problems when used with other Bluetooth devices (Cell phone for example).

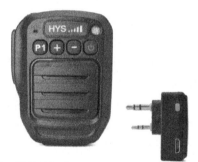

The HYS Wireless Bluetooth Speaker Mic

Radio Holster
Your Baofeng radio comes with a belt clip for attaching to your waste. But – These do break at times, so use of a good belt holster could be a good radio upgrade. I have gone exclusively to using Velcro style radio holsters to keep track of my radios when wearing on the hip.

So, You Bought a Baofeng Radio – Now what?

Cradle & USB Power Cable

Your radio comes with a desktop cradle for using to charge your radio. While this works well, when traveling in the vehicle or out camping, lack of having a 120v power source will keep you from recharging your radio. One option is simply to carry spare batteries. But the Baofeng radios can also be powered from a USB to cradle power adapter allowing you to recharge from any USB/DC Power source.

Additionally, some styles of the Baofeng high-capacity batteries come with a plug and a USB cord allowing you to plug the radio directly into a USB outlet. This is great – removing the cradle from being needed. In my Baofeng two-radio Repeater build, I am opting to use these types of batteries which can then be plugged into a small 12-volt battery with a 12-volt USB outlet, providing the ability to power the repeater off batteries for days, or allowing a 12-volt Solar recharging cell to be connected for even longer use.

AAA/AA Battery Packs

A battery pack using AA or AAA Batteries will allow you to be able to power your radio from conveniently available batteries if necessary. Though this is a good option, my personal preference is still from spare Lithium batteries.

Spare Batteries

One to the top items for your list for your radio will be a spare battery. And not just any spare battery – Go for a Long-life 3800 mAh battery. Mirkit sells these batteries for both the BF-F8HP/UV-5r style radio and the UV-82 series radios with the port for the USB charging cable. This allows easy recharging while on the road.

So, You Bought a Baofeng Radio – Now what?

Other Accessories

In Appendix I in this book we will list sources for many of the accessories that we have listed her. In addition, on our website on our Accessories page we will keep a running list of popular items that you can get for your radio. Many of these will be available on Amazon as well as other sources.

http://www.SoYouBoughtABaofeng.com/Accessories

So, You Bought a Baofeng Radio – Now what?

So, You Bought a Baofeng Radio – Now what?

Appendix A: Calling Frequencies

The calling frequencies that you will use with your Baofeng radio may be one of the following:

2-Meters	144 MHz thru 148 MHz	146.52 MHz
1.25-Meters	222 MHz thru 225 MHz	223.50 MHz
70-Centimeters	420 MHz thru 450 MHz	446.00 MHz

Appendix B: Bands for LEGAL Use

The following are the frequency ranges in each of the bands that Baofeng radios can transmit on that are available for legal Ham radio use:

2-Meters	144 - 148 MHz
1.25-Meters	222 - 225 MHz
70-Centimeters	420 - 450 MHz

Note: The 1.25-Meter band is only available on limited models of Baofeng radios.

Appendix C: GMRS/FRS Frequencies & Data

Following are FRS/GMRS Combined Frequency to Channel information including Power levels and Bandwidths.

USING THE BAOFENG RADIO FOR TRANSMITTING ON THESE FREQUENCIES IS NOT LEGAL.

You should only use an approved radio which carries an FCC Part 95 Certification for GMRS use when transmitting on these frequencies. Also note the minimum power level for transmitting on a Baofeng Radio is 1-watt which exceeds the FRS allowable power on several channels.

Chan	Freq	FRS-PWR	FRS-BW	GMRS-PWR	GMRS-BW	Notes
1	462.5625	2 W	12.5 kHz	5 W	20 kHz	A
2	462.5875	2 W	12.5 kHz	5 W	20 kHz	A
3	462.6125	2 W	12.5 kHz	5 W	20 kHz	A
4	462.6375	2 W	12.5 kHz	5 W	20 kHz	A
5	462.6625	2 W	12.5 kHz	5 W	20 kHz	A
6	462.6875	2 W	12.5 kHz	5 W	20 kHz	A
7	462.7125	2 W	12.5 kHz	5 W	20 kHz	A
8	467.5625	0.5 W	12.5 kHz	0.5 W	12.5 kHz	A
9	467.5875	0.5 W	12.5 kHz	0.5 W	12.5 kHz	A
10	467.6125	0.5 W	12.5 kHz	0.5 W	12.5 kHz	A
11	467.6375	0.5 W	12.5 kHz	0.5 W	12.5 kHz	A
12	467.6625	0.5 W	12.5 kHz	0.5 W	12.5 kHz	A
13	467.6875	0.5 W	12.5 kHz	0.5 W	12.5 kHz	A
14	467.7125	0.5 W	12.5 kHz	0.5 W	12.5 kHz	A
15	462.5500	2 W	12.5 kHz	50 W	20 kHz	A, B
16	462.5750	2 W	12.5 kHz	50 W	20 kHz	A, B
17	462.6000	2 W	12.5 kHz	50 W	20 kHz	A, B
18	462.6250	2 W	12.5 kHz	50 W	20 kHz	A, B
19	462.6500	2 W	12.5 kHz	50 W	20 kHz	A, B
20	462.6750	2 W	12.5 kHz	50 W	20 kHz	A, B
21	462.7000	2 W	12.5 kHz	50 W	20 kHz	A, B
22	462.7250	2 W	12.5 kHz	50 W	20 kHz	A, B

Notes:
A: These channels and frequencies are SHARED by FRS & GMRS radio services
B: This Channel/Frequency is used for GMRS Repeater Output as well.

Appendix D: MURS Frequencies

USING THE BAOFENG RADIO FOR TRANSMITTING ON THESE FREQUENCIES IS NOT LEGAL.
You should only use an approved radio which carries an FCC Part 95 Certification for GMRS use when transmitting on these frequencies. Also note the minimum power level for transmitting on a Baofeng Radio is 1-watt which exceeds the FRS allowable power on several channels.

Channel	Frequency	Maximum bandwidth	Channel name
1	151.82 MHz	11.25 kHz	MURS 1
2	151.88 MHz	11.25 kHz	MURS 2
3	151.94 MHz	11.25 kHz	MURS 3
4	154.57 MHz	20.00 kHz	Blue Dot
5	154.60 MHz	20.00 kHz	Green Dot

- Channels 1–3 must use "narrowband" frequency modulation (2.5 kHz deviation; 11.25 kHz bandwidth).

- Channels 4 and 5 may use either "wideband" FM (5 kHz deviation; 20 kHz bandwidth) or "narrowband" FM.

- All five channels may use amplitude modulation with a bandwidth up to 8 kHz

MURS falls under Part 95 and was not mandated for narrow banding, such as those of Part 90 in the public service bands by January 2013.

So, You Bought a Baofeng Radio – Now what?

MURS is licensed by rule. This means an individual license is not required for an entity to operate a MURS transmitter if it is not a representative of a foreign government and if it uses the transmitter in accordance with the MURS rules outlined in 47 C.F.R Part 95 Subpart J. There is no age restriction regarding who may operate an MURS transmitter.

If you decide that you want to bring MURS communications in as part of your radio communications solution, consider the Baofeng MURS-V1 by Baofeng Tech. This radio is fully compliant and legal for use on MURS whereas the Baofeng BF-F8HP, UV-5R Series, and the UV-82x series are not. The MURS-V1 and the GMRS-V1 radios by Baofeng Tech can be programmed to receive as the BF-F8HP and the UV-82 series radios

Appendix E: VHF Marine Frequencies

USING THE BAOFENG RADIO FOR TRANSMITTING ON THESE FREQUENCIES IS NOT LEGAL.

You should only use an approved radio which carries an FCC Part 95 Certification for GMRS use when transmitting on these frequencies. Also note the minimum power level for transmitting on a Baofeng Radio is 1-watt which exceeds the FRS allowable power on several channels.

Ship XMIT	Ship RECV	Frequency Use
156.050	156.050	Port Operations and Commercial, VTS. Available only in New Orleans / Lower Mississippi area.
156.250	156.250	Port Operations or VTS in the Houston, New Orleans and Seattle areas.
156.300	156.300	Intership Safety
156.350	156.350	Commercial. VDSMS
156.400	156.400	Commercial (Intership only). VDSMS
156.450	156.450	Boater Calling. Commercial and Non-Commercial. VDSMS
156.500	156.500	Commercial. VDSMS
156.550	156.550	Commercial. VTS in selected areas. VDSMS
156.600	156.600	Port Operations. VTS in selected areas.
156.650	156.650	Intership Navigation Safety (Bridge-to-bridge). Ships >20m length maintain a listening watch on this channel in US waters.
156.700	156.700	Port Operations. VTS in selected areas.
--	156.750	Environmental (Receive only). Used by Class C EPIRBs.
156.800	156.800	International Distress, Safety and Calling. Ships required to carry radio, USCG, and most coastal stations maintain a listening watch on this channel.
156.850	156.850	State & local govt maritime control
156.900	156.900	Commercial. VDSMS
156.950	156.950	Commercial. VDSMS
157.000	161.600	Port Operations (duplex)
157.000	157.000	Port Operations
157.050	157.050	U.S. Coast Guard only
157.100	157.100	Coast Guard Liaison and Maritime Safety Information Broadcasts. Broadcasts announced on channel 16.

So, You Bought a Baofeng Radio – Now what?

157.150	157.150	U.S. Coast Guard only
157.200	161.800	Public Correspondence (Marine Operator). VDSMS
157.250	161.850	Public Correspondence (Marine Operator). VDSMS
157.300	161.900	Public Correspondence (Marine Operator). VDSMS
157.350	161.950	Public Correspondence (Marine Operator). VDSMS
157.400	162.000	Public Correspondence (Marine Operator). VDSMS
156.175	156.175	Port Operations and Commercial, VTS. Available only in New Orleans / Lower Mississippi area.
156.275	156.275	Port Operations
156.325	156.325	Port Operations
156.375	156.375	Commercial. Used for Bridge-to-bridge communications in lower Mississippi River. Intership only.
156.425	156.425	Non-Commercial. VDSMS
156.475	156.475	Non-Commercial. VDSMS
156.525	156.525	Digital Selective Calling (voice communications not allowed)
156.575	156.575	Non-Commercial. VDSMS
156.625	156.625	Non-Commercial (Intership only). VDSMS
156.675	156.675	Port Operations
156.725	156.725	Port Operations
156.875	156.875	Port Operations (Intership only)
156.925	156.925	Non-Commercial. VDSMS
156.975	156.975	Commercial. Non-Commercial in Great Lakes only. VDSMS
157.025	157.025	Commercial. Non-Commercial in Great Lakes only. VDSMS
157.075	157.075	U.S. Government only - Environmental protection operations.
157.125	157.125	U.S. Government only
157.175	157.175	U.S. Coast Guard only
157.225	161.825	Public Correspondence (Marine Operator). VDSMS
157.275	161.875	Public Correspondence (Marine Operator). VDSMS
157.325	161.925	Public Correspondence (Marine Operator). VDSMS
157.375	157.375	Public Correspondence (Marine Operator). VDSMS
157.425	157.425	Commercial, Intership only. VDSMS
161.975	161.975	Automatic Identification System (AIS)
162.025	162.025	Automatic Identification System (AIS)

FOR VHF Marine radio use, only use an approved radio for these frequencies. Also remember that many Marine-quality VHF Radios also are water resistant or waterproof where the Baofeng radio is not.

Appendix F: Frequency Coordinators

If you use a HAM Radio for two-way communicating (License required) or just use it to listen in on for conversation and news information, then having a printed reference to the local repeaters in your area and the band plan for your region or state is critical in the event of an emergency that results in power and/or communications and cell system loss. The national frequencies we have provided in this book are great reference sources for general emergency and government frequencies that you can monitor during such an event, but each region or state has lists of repeaters and radio frequencies that have been designated for types of use by organizations that are far too in depth and change too frequently for us to update in this book.

By visiting the Frequency Coordinator organizations for your own state or region you should be able to obtain the plans that are specific to your area. We recommend you go to the appropriate website to find and print those lists and keep them in a folder with your emergency supplies. The list updates frequently enough that an annual refresh of your list should be sufficient. If you are or become an active Ham radio participant, you will become familiar with some of the repeater networks and emergency communication networks in your area. Otherwise having the printed frequencies and contacts for your area will be even more important. For some states or regions no websites were available however specific named contacts with email addresses are sometimes available from the National Frequency Coordinators Council's (NFCC) own website.

So, You Bought a Baofeng Radio – Now what?

National Frequency Coordinators Council - http://www.nfcc.us/

State	Council Name
Alabama	Alabama Repeater Council, Inc http://www.alabamarepeatercouncil.org/
Alaska	Alaska Repeater Coordinators No Web Address
Arizona	Amateur Radio Council of Arizona http://www.azfreqcoord.org/
Arkansas	Arkansas Repeater Council http://www.arkansasrepeatercouncil.org/
California	The Northern Amateur Relay Council of California (NARCC)
California	Southern California Repeater & Remote Base Association
California	Two Meter Area Spectrum Management Association http://www.tasma.org/
California	220 MHz Spectrum Management Association of Southern California http://www.220sma.org/
Colorado	Colorado Council of Amateur Radio Clubs http://www.ccarc.net/
Connecticut	Connecticut Spectrum Management Association, Inc. http://www.ctspectrum.com/
Delaware	The Mid-Atlantic Repeater Council, Inc. (T-MARC) http://www.t-marc.org/
District of Columbia	The Mid-Atlantic Repeater Council, Inc. (T-MARC) http://www.t-marc.org/
Florida	Florida Repeater Council http://florida-repeaters.org/
Georgia	South Eastern Repeater Association, Inc. (SERA) http://sera.org/index.php/home/sera-districts/sera-districts-ga/
Hawaii	HAWAII - Hawaii State Repeater Advisory Council (HSRAC) No Web Address
Idaho	SEICC No Web Address

So, You Bought a Baofeng Radio – Now what?

State	Council Name
Illinois	Illinois Repeater Association, Inc. (IRA) http://www.ilra.net/
Indiana	Indiana Repeater Council http://www.ircinc.org/
Iowa	Iowa Repeater Council http://www.iowarepeater.org/
Kansas	South Eastern Repeater Association, Inc. (SERA) http://www.ksrepeater.com/
Kentucky	South Eastern Repeater Association, Inc. (SERA) http://sera.org/index.php/home/sera-districts/sera-districts-ky/
Louisiana	LCARC- Louisiana Council of Amateur Radio Clubs http://www.lacouncil.net/
Maine	New England Spectrum Management Council (NESMC) http://www.nesmc.org/
Maryland	The Mid-Atlantic Repeater Council, Inc. (T-MARC) http://www.tmarc.org/
Massachusetts	New England Spectrum Management Council (NESMC) http://www.nesmc.org/
Michigan	Michigan Area Repeater Council (MARC) http://www.qsl.net/miarc/
Minnesota	Minnesota Repeater Council (MRC) http://www.mrc.gen.mn.us/
Mississippi	South Eastern Repeater Association, Inc. (SERA) http://sera.org/index.php/home/sera-districts/sera-districts-mi/
Missouri	Missouri Repeater Council, Inc. (MRC) http://www.missourirepeater.org/
Nevada	Northern Nevada - CARCON http://www.carcon.org/
Nevada	Southern Nevada - Southern Nevada Repeater Council (SNRC)

So, You Bought a Baofeng Radio – Now what?

State	Council Name
New Hampshire	New England Spectrum Management Council (NESMC) http://www.nesmc.org/
New Jersey	Metropolitan Coordination Association, Inc. (MetroCor)
New Jersey	Southern/Western New Jersey http://www.arcc-inc.org/
New Mexico	New Mexico Frequency Coordination Committee (NMFCC) No Web Address
New York	NY City & Long Island - Metro Coordination Association, Inc. http://www.metrocor.net/
New York	NE New York - VT Independent Repeater Coordinating Committee http://www.ranv.org/rptr.html
New York	North Central New York - Saint Lawrence Valley Repeater Council http://www.slvrc.org/
New York	Upstate (Mid) New York http://www.unyrepco.org/
New York	Western New York http://www.wnysorc.org/
North Carolina	South Eastern Repeater Association, Inc. (SERA) http://sera.org/index.php/home/sera-districts/sera-districts-nc/
South Carolina	South Eastern Repeater Association, Inc. (SERA) http://sera.org/index.php/home/sera-districts/sera-districts-sc/
Tennessee	South Eastern Repeater Association, Inc. (SERA) http://sera.org/index.php/home/sera-districts/sera-districts-tn/
Virginia	South Eastern Repeater Association, Inc. (SERA)

So, You Bought a Baofeng Radio – Now what?

State	Council Name
	http://sera.org/index.php/home/sera-districts/sera-districts-va/
West Virginia	South Eastern Repeater Association, Inc. (SERA) http://sera.org/index.php/sera-districts-wv/
Ohio	Ohio Area Repeater Council (OARC) http://www.oarc.com/
Oklahoma	Oklahoma Repeater Society Inc. (ORSI) http://www.oklahomarepeatersociety.org/
Oregon	Oregon Region Relay Council, Inc. (ORRC) http://www.orrc.org/
Pennsylvania	Area Repeater Coordination Council, Inc. (ARCC) http://www.arcc-inc.org/
Pennsylvania	Western - Western Pennsylvania Repeater Council (WPRC) http://www.qsl.net/wprc
Rhode Island	New England Spectrum Management Council (NESMC) http://www.nesmc.org/
Texas	Texas VHF-FM Society (TVHFS) http://www.txvhffm.org/
Utah	UVHFS http://www.ussc.com/~uvhfs/
Vermont	VT Independent Repeater Coordinating Committee (VIRCC) - No Web Address
Washington	W. Washington Amateur Relay Association, Inc. (WWARA) http://www.wwara.org/
Wisconsin	Wisconsin Association of Repeaters (WAR) http://www.wi-repeaters.org/
Wyoming	Wyoming Council of Radio Clubs (WCARC) http://www.n7gt.com/coordination.html

So, You Bought a Baofeng Radio – Now what?

Appendix G: Website References

Following are websites that we have used for reference in this book, or sites that we believe will be of strong interest to you.

- Website – http://www.SoYouBoughtABaofeng.cm/
- Book - Gordon West 2018-2022 Technician Class (Yellow Book)
- Book – The ARRL Ham Radio License Manual (Red Book)
- Book – So you Bought A Baofeng?
  http://www.SoYouBoughtABaofeng.com/

- Classes – ARRL Website
  https://www.arl.org/find-an-amateur-radio-license-class/
- Gordon West Radio School
  http://www.GordonWestRadioSchool.com/
- Practice – Ham Radio Prep
  http://www.HamRadioPrep.com/
- Ham Exam
  http://www.HamExam.org/
- Ham Study
  http://www.HamStudy.org/
- Video – Repeater 101 for new Amateur Radio Operators
  https://www.youtube.com/watch?v=TNxyfRDpwpA
- Paper - Understanding SWR by Example
  https://www.arrl.org/files/file/Technology/tis/info/pdq/q1106037.pdf
- Video – Ham Radio Crash Course – Baofeng UV5R Programming
  https://www.youtube.com/watch?v=tyHwAq7w9QE&t=19s/
- A New Hams Guide to Repeaters
  https://www.fmarc.net/license-education/a-new-hams-guide/

Appendix H: SWR Meters

SWR is a topic that is not necessary for normal use of your Baofeng radio. If you do opt to attach an external antenna such as a Car mount/Magnetic mount antenna, or an outdoor antenna on a pole, then SWR is something you should be aware of. Because it is not a primary topic for normal handheld radio use.

SWR (Standing Wave Ratio) is the measurement of power transmitted from a radio that is reflected from the Antenna and through the feed line. This is an important measurement (SWR) that is used for ham radio because it can affect the efficiency of the radio's ability to transmit through a feed line and antenna. SWR readings are taken through a special meter called naturally an SWR Meter. This meter is connected between the radio transmitter and the antenna to measure both the SWR Ratio of power and can measure the power output from the radio.

For the most part, your Baofeng radio won't need to worry about this when working with the stock antenna or with an antenna that is designed for a handheld antenna. If you do decide to work with external antennas connected with Coaxial feed line then using a SWR Meter could be beneficial to ensure you are getting the best signal and efficiency of your antenna.

To understand the SWR meter let's look at the two measurements the meter makes for us.

Standing Wave Ratio (SWR)

This measurement is a ratio measurement of the reflected power back from the antenna. The best reading that we want if possible is a 1:1 ratio. This is read as a two-number value – the higher the number of the left goes the worse the performance you will find from your antenna. Higher ratios will result in a greater percentage of reflected power and voltage back to the transmitter.

So, You Bought a Baofeng Radio – Now what?

In the table shown below we can see that as the ratio increases from 1.0:1 up, the amount of reflected power as well as the % of the voltage reflected rises resulting in a loss in the efficiency of the antenna which means an increase in the loss of your signal out from the radio.

SWR Value	% Power Reflected	% Voltage Reflected
1.0:1	0.00	0
1.1:1	0.20	5
1.2:1	0.80	9
1.3:1	1.70	13
1.4:1	2.80	17
1.5:1	4.00	20
1.6:1	5.30	23
1.7:1	6.70	26
1.8:1	8.20	29
1.9:1	9.60	31
2.0:1	11.00	33
2.5:1	18.40	43
3.0:1	25.00	50
4.0:1	36.00	56
5.0:1	44.40	67
10.0:1	67.00	82

Measuring Power

The other useful feature of one of these meters is the ability to read the amount of power that your radio is putting out. I connected my 5-Watt UV-5r radio to the TX port of both of my SWR Meters with a coax cable and connected a dummy-load on the Antenna side. This dummy load allows me to "Transmit" at full power without a signal being broadcasted. My first reading on both meters read only 4-watts instead of 5. I then replaced the 6' RG-58 Coax cable with a adapter to go directly into the radio – and then the next reading was actually 5-watts. The true reading of the radio. This told me that the 6' RG-58 cable was creating a power loss for me in these measurements.

So, You Bought a Baofeng Radio – Now what?

Digital SWR Meter   Analog SWR Meter

Connecting to a SWR Meter
Connecting a handheld radio can be a little tricky – These meters are generally designed to with a connect with a UHF antenna cable connector (SO239 connector or a UHF PL239 connector). The meters may come with some adapters to facilitate connections, but chances are that you may need either an SMA Female to UHF Female connector, or possibly other adapters depending on your antenna base types. (I have several BNC antennas and need additional adapters).

As mentioned above, you likely will not need a SWR Meter. But you can pick up an analog as little as about $30 which can be useful if you do decide you need one, or if you join with a local Ham club you likely could borrow one from another member. If you do decide to pick one up be watchful to get one rated for the power of your radios, and that works in the proper frequency range. SWR Meters are also popular for CB Radio testing, but a meter designed for CB Radio may work in a different frequency range and may not work with your Baofeng VHF/UHF frequency bands.

Appendix I: Parts Lists, Links to parts, Sources
References to parts and items discussed in this book

Radios:
- Baofeng / BTECH BF-F8HP Radio 8-Watt Radio
  https://baofengtech.com/uv-82hp
- Baofeng / BTECH UV-82HP Radio 8-Watt Radio
  https://baofengtech.com/bf-f8hp
- Baofeng / BTECH UV-5X3 Tri-Band 5-Watt Radio
  https://baofengtech.com/uv-5x3

Radio Distributors / Sources
- Baofeng Tech, US
  https://baofengtech.com/
- Baofeng Radio US (Limited radio options)
  https://baofengradio.us/
- Radioddity
  https://www.radioddity.com/

Antennas:
- Nagoya, Mag-Mount, and other compatible antennas sold through Baofeng Tech
  https://baofengtech.com/accessories
- Signal Stuff Antennas (Signal Stick Antenna)
  https://www.SignalStuff.com
- Abbree Folding whip Antennas
  https://www.amazon.com/42-5-Inch-ABBREE-SMA-Female-Foldable-Tactical/dp/B07STB8BD6

USB Cables:
- BTech PC-03 FTDI Cable
  https://baofengtech.com/Programming-cable

Software:
- CHIRP Programming Software
  https://chirp.danplanet.com/projects/chirp/wiki/Download

So, You Bought a Baofeng Radio – Now what?

- Baofeng Programming Software
  https://baofengtech.com/download
- Programming Image
  http://www.SoYouBoughtABaofeng.com/

Manuals:
- Baofeng Manuals
  https://baofengtech.com/support#manual

Cables/Connectors/Adapters
- SMA-Female to SO-239 Female Pigtail Adapter
  http://www.SoYouBoughtABaofeng.com/Pigtail/
- SMA-UHF RF Connector Kit
  http://www.SoYouBoughtABaofeng.com/SMAKIt/
- SMA Female to BNC Female Adapter
  http://www.SoYouBoughtABaofeng.com/SMAToBNC/
- BNC to UHF – 4 Piece/4-Type RF Connector Kit
  http://www.SoYouBoughtABaofeng.com/BNCUHFKIt/
- PL259 UHF Male t BNC Right-Angle Adapter
  http://www.SoYouBoughtABaofeng.com/PL259ToBNCRightAngle/
- RG-58 15 Meter Cable
  http://www.SoYouBoughtABaofeng.com/RG58_15Meters/
- SMA Female to UHF 3' Cable
  http://www.SoYouBoughtABaofeng.com/SMAUHF3Foot/
- SMA Female to UHF Female Pigtail & Adapter Combo Kit (2 Pieces)
  https://www.amazon.com/dp/B07LF5DCND/ref=cm_sw_em_r_mt_dp_gScmFb57C4Y0

Other Items:
- 23' Telescoping Painter's Pole
  https://www.homedepot.com/p/Mr-Longarm-Pro-Lok-23-ft-Adjustable-3-Section-Extension-Pole-2324/100177392
- HYS Wireless Bluetooth Microphone
  http://www.SoYouBoughtABaofeng.com/BluetoothMic/

Appendix J: CTCSS Squelch Tones (Hz)

The following are the standard set of CTCSS Tones used for repeaters:

67.0	82.5	100.0	123.0	151.4	186.2	225.7
69.3	85.4	103.5	127.3	156.7	192.8	229.1
79.9	88.5	107.2	131.8	162.2	203.5	233.6
74.4	91.5	110.9	136.5	167.9	206.5	241.8
77.0	94.8	114.8	141.3	173.8	210.7	250.3
79.7	97.4	118.8	146.2	179.9	218.1	254.1

Appendix K: DCS Codes

1	D023N	22	D131N	43	D251N	64	D371N	85	D532N
2	D025N	23	D132N	44	D252N	65	D411N	86	D546N
3	D026N	24	D134N	45	D255N	66	D412N	87	D656N
4	D031N	25	D143N	46	D261N	67	D413N	88	D606N
5	D032N	26	D145N	47	D263N	68	D423N	89	D612N
6	D036N	27	D152N	48	D265N	69	D431N	90	D624N
7	D043N	28	D155N	49	D266N	70	D432N	91	D627N
8	D047N	29	D156N	50	D271N	71	D445N	92	D631N
9	D051N	30	D162N	51	D274N	72	D446N	93	D632N
10	D053N	31	D165N	52	D306N	73	D452N	94	D645N
11	D054N	32	D172N	53	D311N	74	D454N	95	D654N
12	D065N	33	D174N	54	D315N	75	D455N	96	D662N
13	D071N	34	D205N	55	D325N	76	D462N	97	D664N
14	D072N	35	D212N	56	D331N	77	D464N	98	D703N
15	D073N	36	D223N	57	D332N	78	D465N	99	D712N
16	D074N	37	D225N	58	D343N	79	D466N	100	D723N
17	D114N	38	D226N	59	D346N	80	D503N	101	D731N
18	D115N	39	D243N	60	D351N	81	D506N	102	D732N
19	D116N	40	D244N	61	D356N	82	D516N	103	D734N
20	D122N	41	D245N	62	D364N	83	D523N	104	D743N
21	D125N	42	D246N	63	D365N	84	D526N	105	D754N

So, You Bought a Baofeng Radio – Now what?

Appendix L: Phonetic Alphabet

Following is the NATO Standard Phonetic alphabet used when communicating call signs across the radio. Use of this alphabet makes it easier to understand characters where individually spoken multiple characters may sound very similar.

Letter	NATO	Letter	NATO
A	Alpha	N	November
B	Bravo	O	Oscar
C	Charlie	P	Papa
D	Delta	Q	Quebec
E	Echo	R	Romeo
F	Foxtrot	S	Sierra
G	Golf	T	Tango
H	Hotel	U	Uniform
I	India	V	Victor
J	Juliet	W	Whiskey
K	Kilo	X	X-ray
L	Lima	Y	Yankee
M	Mike	Z	Zulu

Appendix M: Radio Horizon Antenna Heights

The following is a reference running from 6' to 1000' and the respective ranges. Notice 6' thru 10' are in 2' increments, then 5' to 100' then 50'.

Antenna Height (Ft)	Range (KM)	Range (Miles)
6	5.57	3.46
8	6.43	4.00
10	7.19	4.47
15	8.81	5.47
20	10.17	6.32
25	11.37	7.07
30	12.46	7.74
35	13.46	8.36
40	14.39	8.94
45	15.26	9.48
50	16.09	10.00
55	16.87	10.48
60	17.62	10.95
65	18.34	11.40
70	19.03	11.83
75	19.70	12.24
80	20.35	12.64
85	20.97	13.03
90	21.58	13.41
95	22.17	13.78
100	22.75	14.14
150	27.86	17.31
200	32.17	19.99
250	35.97	22.35
300	39.40	24.48
350	42.56	26.45
400	45.50	28.27
450	48.26	29.99
500	50.87	31.61
550	53.35	33.15
600	55.72	34.62
650	58.00	36.04
700	60.19	37.40
750	62.30	38.71
800	64.34	39.98
850	66.32	41.21
900	68.25	42.41
950	70.12	43.57
1000	71.94	44.70

So, You Bought a Baofeng Radio – Now what?

Appendix N: Weather Radio Frequencies

Not all radio weather channels will come in at all locations. Depending on your location you may receive one or more but not likely all channels. You will need to check your area on NOAA's website for coverage details which is at this website location: http://www.nws.noaa.gov/nwr/

If programming these channels into your radio make sure you have the settings to NOT transmit.

Frequency	Description	Alpha Tag	Band	Mode
162.4000	Weather Radio 1	NWS	Government	FM
162.4250	Weather Radio 2	NWS	Government	FM
162.4500	Weather Radio 3	NWS	Government	FM
162.4750	Weather Radio 4	NWS	Government	FM
162.5000	Weather Radio 5	NWS	Government	FM
162.5250	Weather Radio 6	NWS	Government	FM
162.5500	Weather Radio 7	NWS	Government	FM

Appendix N: Baofeng Menu Settings/Values

Menu#	Menu Setting	Description	Values
0	SQL - Squelch Level	The Squelch level mutes the radio speaker if there is no strong signal. Setting values range from 0 (Off) up through 9. The higher the level of squelch the stronger the signal will have to be to be heard.	0 to 9 where [0] = OFF [1] = Min Sqlch Thru [9] = Max Sqlch
1	Step - Frequency Step	This step value is the frequency step size in VFO/Frequency mode when the radio is scanning or pressing the [UP] and [DN] keys.	[0] - 2.5K [1] - 5.0K [2] - 6.25K [3] - 10.0K [4] - 12.5K [5] - 20.0K [6] - 25.0K [7] | 50.0K
2	TXP - Transmit Power	Allows to select between LOW, MID and HIGH transmitter power.	[0] - HIGH [1] - MID [2] - LOW
3	SAVE - Battery Save	This setting allows the ratio of sleep cycles to wake cycles be adjusted from 1:2, 2:1, 3:1 or 4:1. The higher value extends the battery life the most. When turned on, there is possibility of missing a portion of the received transmission when the monitored frequency activates.	[0] = OFF [1] [2] [3] [4]
4	VOX - Voice Operated Transmit	When turned on the microphone will activate automatically based on set sensitivity.	[0] = OFF And Values [1] thru [10]
5	WN - Wideband/Narrowband	Controls the bandwidth setting of 12.5 kHz or 25 kHz.	[0] = Wide [1] = NARROW
6	6 ABR - Display illumination time	Sets the timeout for the LCD Backlit screen in seconds.	[0] = OFF And then vals |1] thru [10]

So, You Bought a Baofeng Radio – Now what?

7	TDR - Dual Watch, Dual Reception	Sets the dual watch to monitory channel A and Channel B at the same time.	[0] = OFF; [1] = ON;
8	BEEP - Keypad Beep	Turns on or off the Keypress beep.	[0] = OFF; [1] = ON;
9	TOT - Transmission Time-out-Timer	This setting turns off the transmitter after a set time as a way to prevent long transmissions and save battery life. This also helps to prevent transmitting if your PTT button becomes stuck.	15[0] - 600[39] in 15 second steps (TIMEOUT-15)/15=[n]
10	R-DCS Receiver DCS	This option mutes the speaker of the transceiver unless a specific low-level digital signal is received. Without the signal coming though you will not hear anything.	OFF [0] then see DCS Table in Appendix for other values
11	R-CTCS - Receiver CTCSS	This option mutes the speaker of the radio if there is not a specific and continuous sub-audible signal coming through. The station you are listening to MUST transmit this signal for you to hear them.	OFF [0] then see CTCSS Table in Appendix for other values
12	T-DCS - Transmitter DCS	This option allows the transmission of a low-level signal if turned n that will unlock the squelch of a remote radio or repeater.	OFF [0] then see DCS Table in Appendix for other values
13	T-CTCS - Transmitter CTCSS	This option transmits a continuous subaudible signal that will unlock the squelch of a distant receiver which is usually a repeater.	OFF [0] then see CTCSS Table in Appendix for other values
14	VOICE - Voice Prompt	This controls the voice prompt for when a button is pressed. Options are ENGLISH, CHINESE or OFF.	[0] = OFF [1] = ENGLISH [2] = CHINESE
15	ANI-ID - Automatic Number ID	This displays the ANI-ID code set by software - Cannot be changed through the Menu.	Set through Software

16	DTMFST – DTMFSide Tone of transmit code	This option is used to determine when DTMF Side tones can be heard from the radio speaker.	[0]: OFF No DTMF Side Tones are heard DT-ST; [1]: Side Tones heard only from manually keyed DTMF codes ANI-ST; [2]: Side Tones heard only from automatically keyed DTMF codes DT+ANI; [3]: All DTMF Side Tones are heard
17	S-CODE - Signal Code	The S-CODE options selects 1 of 15 DTMF Codes which are programmed with software, and up to 5 digits.	1[0] | 2[1] 3[2] | 4[3] 5[4] | 6[5] 7[6] | 8[9] 9[8] | 10[9] 11[10] | 12[11] 13[12] | 14[13] 15[14]
18	SC-REV - Scanner Resume Method	This sets the method for resuming scanning - Options are "TO", "CO" or "SE"	[0]-TO: Time Operation - scanning will resume after a fixed time has passed; [1]-CO: Carrier Operation - scanning will resume after the signal disappears; [2]-SE: Search Operation - scanning will not resume

So, You Bought a Baofeng Radio – Now what?

19	PTT-ID - When to send the PTT-ID	This determines when to send PTT-ID codes sent during either the beginning or end of transmission. BOT = Beginning of Transmission EOT = End of Transmission Both = Both BOT & EOT	[0]-OFF: No ID is sent; [1]: BOT = S-CODE is sent at beginning; [2]: EOT = Selected S-CODE is sent at the ending; [3]: BOTH = The selected S-CODE is sent at beginning and ending
20	PTT-LT - Signal code sending delay	Sets the delay in milliseconds of the PTT-ID	0 - 50ms
21	MDF-A - Channel Mode A Display	This sets the display mode of channel A. Display modes can be FREQ for Frequency; NAME for channel name (Set through Software only), or CH for Channel Number.	[0]: CH - Displays the channel number; [1]: NAME - Displays the channel name; [2]: FREQ - Displays programmed Frequency
22	MDF-B - Channel Mode B Display	This sets the display mode of channel B. Display modes can be FREQ for Frequency; NAME for channel name (Set through Software only), or CH for Channel Number.	[0]: CH - Displays the channel number; [1]: NAME - Displays the channel name; [2]: FREQ - Displays programmed Frequency
23	BCL - Busy Channel Lock-out	This option will disable the [PTT] button if that channel is already in use. The radio will beep and will not transmit if the channel is in use.	[0] OFF; [1] ON;

So, You Bought a Baofeng Radio – Now what?

24	AUTOLK – Automatic Keypad Lock	Turning this ON locks the keypad if not used in 8 seconds. To unlocal press the [# ⌂O] key for 2 seconds.	[0] OFF; [1] ON;
25	SFT-D - Frequency Shift Direction	Selects the Repeater Frequency Offset Direction (-) or (+)	[0]: OFF - TX = RX (simplex) +; [1]: TX will be shifted higher in frequency than RX -; [2]: TX will be shifted lower in frequency than RX
26	OFFSET - Frequency shift amount	This sets the offset amount between TX and RX frequencies.	00.000 - 69.990 in 10 kHz steps
27	MEM-CH - Store a Memory Channel	Use this option to create a new channel or modify an existing one for access in Channel mode.	000 - 127
28	DEL-CH - Delete a memory channel	This menu option is used to delete the information programmed into an existing channel for reprogramming or leaving empty.	000 - 127
29	WT-LED - Display backlight color, Standby	Default: PURPLE	[0]: OFF; [1] = BLUE; [2] = ORANGE; [3] = PURPLE;
30	RX-LED - Display backlight colorReceive	Default: BLUE	[0]: OFF; [1] = BLUE; [2] = ORANGE; [3] = PURPLE;
31	TX-LED - Display backlight colorTransmit	Default: ORANGE	[0]: OFF; [1] = BLUE; [2] = ORANGE; [3] = PURPLE;

So, You Bought a Baofeng Radio – Now what?

#	Setting	Description	Values
32	AL-MOD - Alarm Mode	You have three options here which are: SITE=This sounds the alarm through the radio speaker only; TONE=This sends a cycling tone transmitting over the air; or CODE=Transmits '119' followed by the ANI code over he air (Reverse of 911)	[0]: SITE - Sounds alarm through your radio speaker only; [1]: TONE - Transmits a cycling tone over-the-air; [2]: CODE - Transmits '119' (911 in reverse?) followed by the ANI code over-the-air;
33	BAND - Band Selection	This sets the [A] or [B] channel to the VHF or UHF band when in VFO/Frequency mode.	[0]: VHF; [1]: UHF;
34	TDR-AB - Transmit selection while in Dual Watch mode	Use this to set the priority to the selected display (A or B) when the signal in the opposing display stops.	[0]: OFF; [0]: A; [1]: B;
35	STE - Squelch Tail Elimination	Use this to eliminate the squelch tail noise that occurs from Baofeng handheld radios that are working in simplex mode (Radio-to-Radio without a repeater).	[0]: OFF; [1]: ON;
36	RP-STE - Squelch Tail Elimination	Use this to eliminate the squelch tail noise when operating through a repeater.	[0]: OFF; Then values 1 – 10
37	RPT-RL - Delay the squelch tail of repeater	This setting delays the tail tone of the repeater	[0]: OFF; Then values 1 – 10
38	PONMSG - Power On Message	This controls how the display acts when the transceiver is turned on.	[0]: FULL - Performs an LCD screen test at power-on; [1]: MSG - Displays a 2-line power-on message

39	ROGER - Roger Beep	This sends an End-of-Transmission tone to indicate to other stations that the transmission message has ended.	[0]: OFF; [1]: ON;
40	RESET - Restore defaults	Performs a factory reset.	[0]: VFO; [1]: ALL;
41 (UV-82) Radio ONLY	R-TONE – Repeater Tone	The R-TONE is short for Repeater Tone and is used to activate repeaters that require a specific audible tone to be transmitted for access. Transmit by pressing the [F] side key while the [PTT] button is also pressed. UV-82x series only.	[0] 1000 HZ; [1] 1450 HZ; [2] 1750 HZ; [3] 2100 HZ;

So, You Bought a Baofeng Radio – Now what?

Appendix O: Chirp Radio Settings Menus (Baofeng Radios)

Settings – Basic Radio Details:

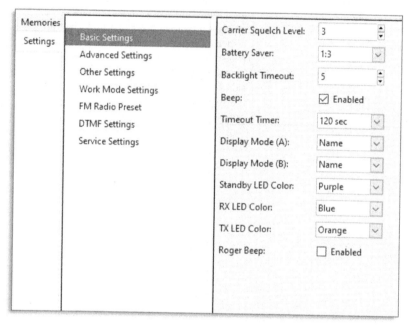

CHIRP Radio Screen 1

So, You Bought a Baofeng Radio – Now what?

Appendix O (Continued)
Settings – Advanced Settings through CHIRP

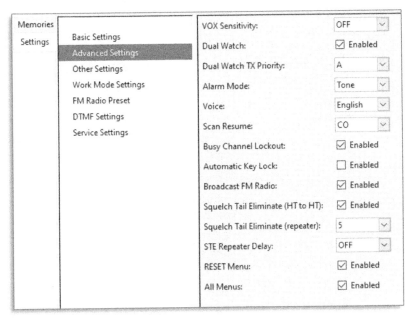

CHIRP Radio Screen 2

So, You Bought a Baofeng Radio – Now what?

Appendix O (Continued)
Settings – Advanced Settings through CHIRP

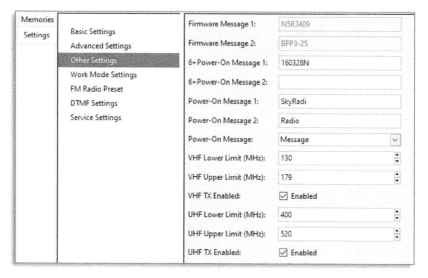

CHIRP Radio Screen 3

So, You Bought a Baofeng Radio – Now what?

Appendix O (Continued)
Settings – Work Mode through CHIRP

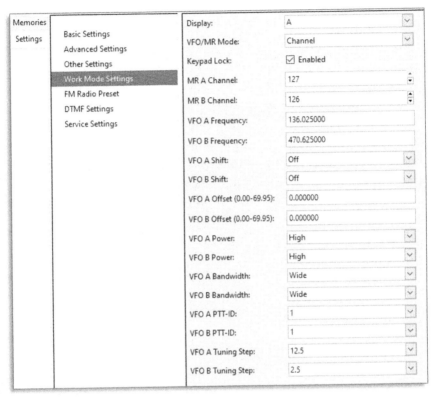

CHIRP Radio Screen 4

So, You Bought a Baofeng Radio – Now what?

Appendix O (Continued)
Settings – FM Radio Preset settings through CHIRP

Memories		FM Preset(MHz):	93.3
Settings	Basic Settings		
	Advanced Settings		
	Other Settings		
	Work Mode Settings		
	FM Radio Preset		
	DTMF Settings		
	Service Settings		

CHIRP Radio Screen 5

So, You Bought a Baofeng Radio – Now what?

Appendix O (Continued)
Settings – DTMF Settings through CHIRP

Memories			
Settings	Basic Settings	PTT ID Code 1:	20202
	Advanced Settings	PTT ID Code 2:	
	Other Settings	PTT ID Code 3:	
	Work Mode Settings	PTT ID Code 4:	
	FM Radio Preset	PTT ID Code 5:	
	DTMF Settings	PTT ID Code 6:	
	Service Settings	PTT ID Code 7:	
		PTT ID Code 8:	
		PTT ID Code 9:	
		PTT ID Code 10:	
		PTT ID Code 11:	
		PTT ID Code 12:	
		PTT ID Code 13:	
		PTT ID Code 14:	
		PTT ID Code 15:	30303
		ANI Code:	80808
		ANI ID:	Off
		Alarm Code:	119
		DTMF Sidetone:	DT+ANI
		DTMF Speed (on):	80 ms

CHIRP Radio Screen 6

So, You Bought a Baofeng Radio – Now what?

Appendix O (Continued)
Settings – Service Settings through CHIRP

Memories / Settings		
	Basic Settings	VHF Squelch 0: 0
	Advanced Settings	VHF Squelch 1: 22
	Other Settings	VHF Squelch 2: 24
	Work Mode Settings	VHF Squelch 3: 26
	FM Radio Preset	VHF Squelch 4: 27
	DTMF Settings	VHF Squelch 5: 29
	Service Settings	VHF Squelch 6: 30
		VHF Squelch 7: 31
		VHF Squelch 8: 32
		VHF Squelch 9: 33
		UHF Squelch 0: 0
		UHF Squelch 1: 14
		UHF Squelch 2: 15
		UHF Squelch 3: 16
		UHF Squelch 4: 17
		UHF Squelch 5: 18
		UHF Squelch 6: 19
		UHF Squelch 7: 20
		UHF Squelch 8: 21
		UHF Squelch 9: 22

CHIRP Radio Screen 7

So, You Bought a Baofeng Radio – Now what?

Made in the USA
Middletown, DE
15 October 2023

40812453R00116